Mycotoxins and Food Safety

Edited by Suna Sabuncuoğlu

Published in London, United Kingdom

IntechOpen

Supporting open minds since 2005

Mycotoxins and Food Safety
http://dx.doi.org/10.5772/intechopen.77743
Edited by Suna Sabuncuoğlu

Contributors
Kobun Rovina, Nurul Shaeera Sulaiman, Joseph Vonnie Merrylin, Anna Abdolshahi, Behdad Shokrollahi Yancheshmeh, Aycan Cinar, Elif Onbaşı, Constanze Pietsch, Jianhua Wang, Zhiyong Zhao, Xianli Yang, Junhua Yang, Andong Gong, Jingya Zhang, Lei Chen, Changyan Zhou, Parise Adadi, Daniel Nsengumuremyi, Nadezhda V. Barakova, Elena F. Krivoshapkina, Gavers K. Oppong, Suna Sabuncuoğlu, Su Xin Yi

First published in London, United Kingdom, 2020 by IntechOpen
IntechOpen is the global imprint of INTECHOPEN LIMITED, registered in England and Wales, registration number: 11086078, 5 Princes Gate Court, London, SW7 2QJ, United Kingdom
Printed in Croatia

British Library Cataloguing-in-Publication Data
A catalogue record for this book is available from the British Library

Additional hard and PDF copies can be obtained from orders@intechopen.com

Mycotoxins and Food Safety
Edited by Suna Sabuncuoğlu
p. cm.
Print ISBN 978-1-78984-874-8
Online ISBN 978-1-78984-875-5
eBook (PDF) ISBN 978-1-83968-894-2

For EU product safety concerns:
IN TECH d.o.o., Prolaz Marije Krucifikse Kozulić 3, 51000 Rijeka, Croatia, info@intechopen.com or visit our website at intechopen.com.

Meet the editor

Suna Sabuncuoğlu graduated from Faculty of Pharmacy, University of Hacettepe, Ankara, Turkey. She completed her PhD studies in Pharmaceutical Toxicology, and as a PhD student she worked at the *International Agency for Research on Cancer, Molecular Carcinogenesis Laboratory.* She also had a postdoc position at the *Department of Chemotherapy and Virology, Rega Institute, Catholic University of Leuven.* Dr. Sabuncuoğlu became a lecturer in 2013 and an associate professor in 2014. To date, she has served on many different boards, commissions and centers in and out of the university. Since 2018, she has had the title of European Registered Toxicologist (ERT). She continues to work as an associate professor at Hacettepe University, Faculty of Pharmacy, Department of Pharmaceutical Toxicology.

Contents

Preface

Foodborne illnesses are a worldwide issue. Many different mycotoxins have been identified, but the most commonly observed include aflatoxins, ochratoxin A, patulin, fumonisins, zearalenone, and nivalenol/deoxynivalenol. Mycotoxins can cause several adverse health effects in living organisms including immunodeficiency and cancer. This book provides information about foodborne mycotoxins, their toxicities, new determination methods, prevention strategies, and regulations. It also describes different food safety strategies, risk assessment, and recent detection techniques such as biosensors and nanoparticles.

Food safety is increasingly viewed as an essential global public health issue. Many countries have collaborated with the World Health Organization (WHO) in order to improve their food safety systems and have updated their national legislation. The WHO encourages national authorities to evaluate accurately the levels of mycotoxins in foodstuff on their market and comply with both national and international maximum levels, conditions, and legislation. Governments play critical roles in protecting the food supply. However, many countries do not have sufficient equipment to respond to existing and emerging food safety problems. In addition, there is limited information available to fully evaluate food safety problems and issues. In this respect, national experiences and knowledge have to be shared.

Dr. Suna Sabuncuoglu
Associate Professor,
Faculty of Pharmacy,
Department of Toxicology,
Hacettepe University,
Sıhhiye, Ankara, Turkey

Chapter 1

Introductory Chapter: Mycotoxins and Food Safety

Suna Sabuncuoğlu

1. Introduction

Food-borne illnesses are prevalent in all parts of the world, and the toll in terms of human life and suffering is enormous. Contaminated food contributes to 1.5 billion cases of diarrhea in children each year, resulting in more than 3 million premature deaths, according to the World Health Organization (WHO). Food safety is used as a scientific discipline describing handling, preparation, and storage of food in ways that prevent food-borne illness. The occurrence of two or more cases of a similar illnesses resulting from the ingestion of a common food is known as a food-borne disease outbreak [1, 2].

Food safety issues can have very different political implications. Understanding the potential for the application of Multi-Criteria Decision Analysis processes in countries with challenges on data availability, limited processes for stakeholder input to decision-making, and so on, is an important foundation for the development of FAO guidance for food safety decision-making using best available evidence for transparent decision-making [3].

Recent research has increased the awareness of chemical residues and natural contaminants in food. At the same time, consumer concerns about food safety have also grown. At a national and international level, this has resulted in more stringent imposition of new, legislative limits for a range of mycotoxins which can contaminate food raw materials and enter the food chain [4].

Mycotoxins are naturally occurring toxins produced by microfungi that are capable of causing disease and death in living organisms. The fungi grow on a variety of different foodstuffs including cereals, nuts, spices, dried fruits, apples and coffee beans, often under warm and humid conditions [5, 6]. It is generally known that cereals, peanuts, spices, coffee, and herbal teas can be contaminated with mycotoxins. Various cereal and crops have potential fungal attack either in the field or during storage [6, 7].

The adverse effect of molds and fungi was known already in ancient times. In the Middle Ages, outbreaks of ergotism caused by ergot alkaloids from *Claviceps purpurea* reached epidemic proportions, mutilating and killing many people in Europe. Some mycotoxicoses have disappeared due to more rigorous hygiene measures such as citreoviridin-related malignant acute cardiac beriberi and alimentary toxic aleukia. General interest in mycotoxins increased in 1960 when a feed-related mycotoxicosis called turkey X disease, which was caused by aflatoxins, appeared in farm animals in England. Subsequently, it was found that aflatoxins are hepatocarcinogens in animals and humans, and this stimulated research on mycotoxins.

Mycotoxins have attracted worldwide attention because these have been recognized as a major economic problem due to the significant economic losses associated with their impact on human health, animal productivity, and domestic and international trade [7].

Mycotoxins are produced by fungi such as Aspergillus, Penicillium, Fusarium, or Alternaria. These fungi may produce as secondary, metabolites a diverse group of chemical substances known as mycotoxins. Several hundred different mycotoxins have been identified, but the most commonly observed mycotoxins that present a concern to human health and livestock include aflatoxins, ochratoxin A, patulin, fumonisins, zearalenone, and nivalenol/deoxynivalenol. It is possible to be wide year to year fluctuations in the levels of mycotoxins in foods [8]. This can be dependent on many factors including adverse conditions favoring fungal invasion and growth. Mycotoxicoses, which can occur in both industrialized and developing countries, arise when environmental, social, and economic conditions combine with meteorological conditions (humidity, temperature) which favor the growth of molds. Factors affecting mycotoxin formation are listed below:

- Plant-related factors (type and sensitivity of the plant, other toxic fungal species found in the plant, water content of the plant, plant maturity, mechanical damage to the plant; for example, damage to the plant by insects and/or birds)

- Environmental factors (temperature and humidity of the environment where the plant grows, oxygen source of the environment)

- Conditions during processing, storage conditions after harvest and during storage (relative humidity and temperature of the environment) [6, 9]

Mycotoxins are toxic secondary metabolites that are synthesized by various types of pathogenic fungi. When they are taken into the organism, they can cause latent, acute, or chronic pathological conditions in humans and animals. With developing the modern farming, storage and processing practices, the aim is to reduce obvious contamination, and much of our concern now focuses on chronic effects at low levels of exposure. Thus, several mycotoxins are potent animal carcinogens and have been classified by the International Agency for Research in Cancer (IARC, 1993) as human carcinogens or potential (probable and possible) human carcinogens [2, 7].

Quality procedures and legislation of levels that are toxicologically acceptable are needed to minimize the exposure to mycotoxins; these actions are carried out in the agricultural practice, storage of products, and control of products intended for human or animal consumption [4]. The techniques used for mycotoxin determination are chromatography, including high-performance liquid chromatography (HPLC), thin-layer chromatography (TLC), and gas chromatography–mass spectrometry (GC–MS), and enzyme-linked immunosorbent assay (ELISA) techniques. Considering the limitations of these techniques, the high cost, lack of sensitivity, and need for a skilled technician, there is an urgent need for other accurate, simple, and cost-effective techniques [10].

This book will provide updated information about food-borne mycotoxins, their toxicities, new determination methods, prevention strategies, and regulations in the world.

Author details

Suna Sabuncuoğlu
Faculty of Pharmacy, Department of Toxicology, Hacettepe University, Sihhiye, Ankara, Turkey

*Address all correspondence to: suna@hacettepe.edu.tr

References

[1] Kumar P et al. Aflatoxins: A global concern for food safety, human health and their management. Frontiers in Microbiology. 2017;7:2170

[2] Toxic effects of mycotoxins in humans. Available from: https://webcache.googleusercontent.com/search?q=cache:4xWocC3OHEEJ:https://www.who.int/bulletin/archives/77(9)754.pdf+&cd=4&hl=tr&ct=clnk&gl=tr&client=firefox-b-d [Last Accessed: 10 June 2020]

[3] The Food and Agriculture Organization, Food safety and quality, Policy Advice. Available from: http://www.fao.org/food/food-safety-quality/capacity-development/policy-advice/en/ [cited 10 June 2020], [Last Accessed: 10 June 2020]

[4] Sanzini E et al. Quality control of plant food supplements. Food & Function. 2011;**2**(12):740-746

[5] Kabak B, Dobson AD. Mycotoxins in spices and herbs—An update. Critical Reviews in Food Science and Nutrition. 2017;**57**(1):18-34

[6] Pitt J et al. Mycotoxins and toxigenic fungi. Medical Mycology. 2000;**38** (Suppl 1):41-46

[7] Marin S et al. Mycotoxins: Occurrence, toxicology, and exposure assessment. Food and Chemical Toxicology. 2013;**60**:218-237

[8] Bennett JW, Klich M. Mycotoxins. Clinical Microbiology Reviews. 2003;**16**(3):497-516

[9] Blesa J et al. Factors affecting the presence of ochratoxin A in wines. Critical Reviews in Food Science and Nutrition. 2006;**46**(6):473-478

[10] Bueno D et al. Determination of mycotoxins in food: A review of bioanalytical to analytical methods. Applied Spectroscopy Reviews. 2015;**50**(9):728-774

Chapter 2

Food Contamination

Anna Abdolshahi and Behdad Shokrollahi Yancheshmeh

Abstract

This chapter discusses food contamination including mycotoxin contamination problems, biological, chemical, physical, and cross contamination. Food contamination challenges are generally referred to the presence of microorganisms or derived toxic substances such as mycotoxin in food that make them unsafe for human, animals, and crops. The mycotoxins can enter food throughout the food supply chain (from farm to fork). In terms of the safety of food, the presence of mycotoxin is a hazard threatening the consumer of contaminated food. Furthermore, it is necessary to know the nature, sources, distribution ways, and incidence of mycotoxin contamination in order to protect people and provide public health.

Keywords: contamination, biological, food, mycotoxin

1. Introduction

Food contamination refers to the ways that food has been depraved biologically, physically, or chemically. The contaminant could inter the food unintentionally pending agricultural production, environment, storage, transportation, sale and processing. In general, two sources of contamination are outside sources and formation in food that refer to primary and secondary contamination respectively [1]. The main criteria for contamination judgment could be potential risk and the effect it has on human health. In this regard, mycotoxins and other microbial toxins, toxic elements, radioactive isotopes, nitroso compound, polycyclic hydrocarbon aromatic, halogen containing organic compounds, pesticides residues, veterinary drug residues, etc., are major critical food contaminants [2].

Mycotoxins produced by filamentous microfungi that can cause many diseases in vertebrate animals via ingestion, absorbtion (through the skin) and inhalation routes. Mycotoxins have been found in a variety of food commodities due to the mycotoxin producer fungus are able to grow on a vast range of foods. The most pronounced contamination has been initiated from the agricultural fields during several harvesting stages including preharvest, harvest, and postharvest. Mycotoxin contaminations finally continue to the consumer table. Poor management in all stages not only can lead to rapid deterioration in nutritional value but also provides proper condition for fungal growth and also mycotoxin production. The most efficient way to control food contaminants is the implementation of Good Manufacturing Practices (GMPs) and Hazard Analysis and Critical Control Point (HACCP) that will help prevent hazards in life [3]. New approaches are based on identification of critical control point in production/processing of food that obtain optimum condition for mycotoxin production. To minimize and control mycotoxins in food chain all environmental and climate factors must be assessed.

This chapter will review a summary of food contamination types including biological, chemical, physical and cross contamination. We will also discuss mycotoxin contamination problems regarding the main stages of food production chain.

2. Biological contamination

Biological contamination generally realizes as contamination of food or environment with microorganisms and their derivatives such as toxins. In this regard, bacteria, viruses, fungi, and parasites are potential contaminants. They are found in food, walls, water, air, clothes, etc. The biological contamination also can occur via macroscopic organisms including rodents and insects. The biological contaminants cause human diseases via three mechanisms including infection, intoxication and immunologic responses [4, 5].

2.1 Bacteria

Bacteria are small microorganisms that can grow in an ideal condition. They split and multiply so quickly [6]. Harmful bacteria, called pathogen, are recognized as hazards in safety of food. Therefore the spread and incidence of them must be controlled in food. The common sources for bacterial growth and further distribution are the air, human body, dust, pets and pests, row food (meat, milk, vegetable, etc.), soil, kitchen/factory instruments, food handlers and cloths/hands. The extrinsic factors that provide optimum conditions for bacteria to survive include food (especially protein), water (water activity), oxygen, temperature, and pH level [7]. The control of these factors can result in well preservation of food [8]. **Table 1** is illustrates major bacteria and their risks.

Major bacteria	Risk contamination
Clostridium botulinum	Intoxication, even death
Listeria monocytogenes	Infection
Salmonella spp. (*typhimurium*, *enteriditis*)	Infection
Enterohaemorrhagic *Escherichia coli*	Infection
Campylobacter jejuni	Infection
Yersinia enterocolitica	Infection
Listeria monocytogenes	Infection
Bacillus anthracis	Infection
Bacillus cereus	Intoxication
Staphylococcus aureus	Intoxication
Clostridium perfringens	Infection
Vibrio spp. (*vulnificus*, *parahaemolyticus*)	Infection
Brucella abortus, *B. suis*	Infection
Shigella spp. (*dysenteriae*)	Infection

Table 1.
The major bacteria and their risks.

2.2 Viruses

Viruses are very tiny organisms that can grow and survive only in a host cell. They are able to enter food and water due to poor hygienic conditions. Viruses can also be found in people who disrespect hygienic practices [9]. Viruses can only multiply and grow inside a living cell. They are very resistant to heat, drying, freezing, radiation, etc., and are also able to survive for a long time in food or environment. Viruses can enter food during processing, transportation through person to person contact [10]. The awareness about the importance of viruses as food contaminants would result in good hygiene practices done by consumers to minimize the transmission of viral illnesses [2]. **Table 2** shows the major viruses and their risk.

2.3 Parasites

Parasite including worms and protozoa can enter food or water. They can also infect people through these ways. They need ambient environment and proper hosts to survive. Contamination of food occurs by fecal due to poor personal hygiene of food handler, improper disposal of human feces, improper sewage treatment and utilization of untreated sewage for crop culturing [3, 11]. **Table 3** shows major Parasites and their risks.

Major viruses	Risk contamination
Hepatitis A virus	Fever, abdominal discomfort
Norwalk virus	Nausea, vomiting, diarrhea, abdominal pain (gastroenteritis), headache, low-grade fever
Rotavirus	Vomiting, watery diarrhea, fever, abdominal pain

Table 2.
The major viruses and their risk.

Major parasites	Risk contamination
Giardia lamblia	Diarrhea, abdominal cramps, fatigue, nausea, flatulence (intestinal gas), weightloss
Entamoeba histolytica	Dysentery (severe, bloody diarrhea)
Ascaris lumbricoides	Intestinal, lung infection
Diphyllobothrium latum	Attaches to intestinal wall
Cryptosporidia	Respiratory, gastrointestinal illness
Trichinella spiralis	Intestinal wall, enter the blood (to feed on it) and lymphatic system
Toxoplasma gondii	Neurological disorders, particularly schizophrenia, bipolar disorder
Taenia solium	Attaches to intestinal wall
Anisakis spp.	Anisakiasis

Table 3.
The major parasites and their risks.

3. Chemical contamination

Chemical contaminations of food are another human concern that involves numerous substances such as: agrochemicals, veterinary medicines residues, pesticides residues, environmental contaminants, heavy metals, persistent organic pollutants, and natural toxins; which happen in food during chemical reactions at processing such as acrylamide, furan, and heterocyclic amines [12]. Other processes leading to the formation of contaminants include fermentation (e.g., ethyl carbamate, 3-monochloropropanediol) and disinfection (e.g., trihalomethanes). On the other hand, food contact materials are also kinds of chemical sources (e.g., formaldehyde, melamine, phthalates, and primary aromatic amines) that are able to leach into food. Some chemicals are naturally present in the environment, which includes ubiquitous pollutants such as dioxins and heavy metals may be increased by anthropogenic activity [13].

Some common sources of chemical contamination are:

- cleaning agents;
- unwashed fruits and vegetables;
- food containers made from non-safe polymers;
- pest control products; and
- chemicals used in equipment maintenance.

4. Physical contamination

Physical contamination refers to food that is contaminated by a foreign object during production process stages. Physical contaminants in food could come from external sources, (such as metal fragments), or internal sources (such as bone particles and pits). They can enter food accidentally during pre- and postharvesting due to poor agricultural practices and also in manufacturing, storage, transportation, or retail. Some physical contaminants are considered as food safety concerns such as glass. Sometimes a physical contaminated food can also be biologically contaminated such as the presence of a fingernail. **Table 4** provides a summary of common sources of physical contaminants in foods [12].

Sources	Contaminants
Field	Rocks/stones/sand, asphalt, metals/bullets, concrete particles, bones, wood fragments and thorns
Processing	Glass, ceramic/chards, metal fragments, staples, blades, clips, needles, keys, screws, magnet fragments, washers, bolts, screening, plastics, grease/lubricants, rubber, insulation/seal materials, nail polish, jewelry, coins, pieces of gloves, finger cots, bandages, cigarette butts, gum, bones, pits, fruit stones, nut and animal shells, medications/tablets/capsules, wood, pens, and pencils, rodents and insects
Storage and distribution	Metal, plastic, and wood fragments, insects and rodents

Table 4.
Common sources of physical contaminants in foods.

5. Cross contamination

Cross-contamination occurs in food due to the contact of a contaminated substance coming from another food via many different ways and various sources including: Dirty clothes, Utensils, Coughing, sneezing or even touching the face and hair with food, Pests, Flies, cockroaches, mice and rats, Contacting Raw food, Garbage and waste materials [12].

6. Mycotoxin contamination problem

Mycotoxins are secondary metabolites that are produced by molds. Mycotoxins belong to biological contamination category. The majority of fungi can produce mycotoxins yet this potential is species specification [14]. Several adverse effects on humans, animals, and crops originate from mycotoxins. The contamination of food with mycotoxins is a worldwide concern. Incidence of mycotoxins depends on temperature and humidity of a region that is prone to the fungal growth. The exposure to mycotoxins could take place by ingestion and or dermal and inhalation ways without involving the producer fungi. In fact these fungal toxins are a kind of abiotic hazard originated from biotic ones [15]. The disease caused by mycotoxins is mycotoxicoses also resulted in different acute and chronic effects [16]. Generally, the contaminants could enter the food unintentionally by agricultural production, environment, storage, transportation, sale and processing. Mycotoxins are natural contaminants contribute the food chain. The mycotoxin may contaminate the food during several stages of food chain from the soil to the plate. As a matter of fact mycotoxin contamination can occur in food by infection of crops not only when directly consumed by human but also consumed as feed. However ingested mycotoxin could result in its accumulation in body organs that enter food/feed through agricultural products, meat, milk or eggs. Various foods such as cereals, nuts, spices, fruits and also their products have a potent to be contaminated with mycotoxins at high content. **Table 5** shows major mycotoxin and producer microorganism.

The foods could contaminate with fungal toxins from farm at post- and preharvest stages. The implementation of Good Agriculture Practice and Good Manufacturing Practice are efficient strategies in preventing of mycotoxin contamination [17]. However, every negligence in this field could provide proper condition for fungal growth and proliferation as well as *Aspergillus* growth and aflatoxin production in nuts [18]. Therefore, the condition of production, processing, drying, handling, storage, transportation, and marketing must be controlled. It should be considered that further mycotoxin increase is difficult if the food products are preserved or stored under preventive conditions for fungal growth and mycotoxin production especially regarding water activity and temperature. The awareness of all people either producer or consumer about the ways of mycotoxin entering the food, the main stages of food chain involving in mycotoxin contamination, the optimum condition for increase of mycotoxin contamination risk and critical control tips in this field are necessary to the prevention of mycotoxin contamination.

6.1 Agricultural production

Mycotoxins contamination of agricultural commodities can initiate from field and obtaining of conditions that conduct the fungal growth. The crops can be infected by molds at any line in the field. The production of mycotoxins due to mold growth is commonly associated with variation in weather conditions, plant stress, and humidity also inadequate feeding conditions [19]. In this regard there are three main stages that develop the mycotoxin contamination in food staff from agricultural aspect. These stages include:

Mycotoxin name	Producer microorganism
Aflatoxins	*A. flavus, A. parasiticus*, *Aspergillus bombycis*, *Aspergillus ochraceoroseus*, *Aspergillus nomius*, and *Aspergillus pseudotamari*
Ochratoxins (ochratoxin A)	*Aspergillus ochraceus*, *Aspergillus carbonarius*, *Aspergillus melleus*, *Aspergillus sclerotiorum*, *Aspergillus sulphureus*, *Pichia verrucossum*
Trichothecenes	*Fusarium* spp.
Zearalenone	*Fusarium*, *F. culmorum*, *F. graminearum*, *F. sporotrichioides*
Fumonisins	*Fusarium proliferatum*, *Fusarium verticillioides*
Tremorgenic toxins	*Penicillium*
Ergot alkaloids	*Claviceps*
Moniliformin	*Fusarium* species (mainly *F. proliferatum*)

Table 5.
The major mycotoxins and producer microorganism.

6.2 Preharvest

Mycotoxins can be produced in some natural food products due to the plants that are infected by mycotoxigenic species of molds group from farm. Preharvest practices include obtaining proper planting conditions such as soil ingredient, field qualification, crop rotation, irrigation, insect prevention, and antifungal treatment [20]. As a matter of fact the preharvest condition control is the first line of mycotoxin prevention therefore the implementation of good agricultural practices (GAP) is needful. Some important tips for preventing mycotoxins in preharvest stage are listed below:

- control of climate;
- control of fertilizer;
- control of insect, rodent, and birds;
- control of weed;
- biological control;
- control of planting date control;
- control of irrigation time;
- prevention of early splitting in nuts; and
- control of water activity and water stress.

6.3 Harvest

Mycotoxin production could intensify by any inconsideration in harvest level. The main strategies in harvesting are including utilization of efficient harvesting/collecting/transportation equipment, attention to harvesting time, control of moisture and full maturity of product, inhibition of crop damages during harvesting. The delayed harvest could influence the development of mycotoxin contamination.

The aflatoxin incidence in maize and nuts during delayed harvest were reported in many literatures. In the harvest stage it should be avoid contacting the harvested crop with the ground in order to prevent further contamination [21]. Some of major tips in order to prevention of mycotoxins in harvest stage are listed below:

- control of date of harvest;
- control of the last irrigation;
- control of storage condition;
- control of blending of harvested products;
- control of the contact with soil/ground;
- control of harvest equipment hygiene;
- control of damaging of grain; and
- sorting of defectives.

6.4 Postharvest

Mycotoxin contamination more likely could occur in the postharvest stage due to improperly handling. The high humidity during postharvest prone the dry seed to absorb moisture followed by increase in water activity that conductive to contamination. Also the combination of temperature with moisture results in the extent of mycotoxin contamination risk [21, 22]. At the postharvest stage, the observation of hygienic in all practices is necessary for safety guarantee of food products for example the use of clean transport vehicle free of any fungal growth. Consequently some tree nuts with high risk of mycotoxin contamination should be transferred to the processing plant as soon as possible after full maturity approximately within 6 h. The time of harvesting strictly have been recommended influence in mycotoxin production. Some crops when left in massive volume on the farm for a long time may present high level of mycotoxins. Some of important tips in postharvest stage in order to prevention of mycotoxins are listed below:

- rapid dehulling of crops (if it is needed);
- rapid drying of crops (if it is needed);
- separation of early splitting grain;
- control of moisture content of product;
- control of time and temperature during processing;
- control of temperature and relative humidity during storage;
- control of hygienic condition in all process;
- control of additional water content of crops after washing;

- control of packaging condition; and
- control of chemical preservatives.

6.5 Environment and climate

Environmental conditions could increase the susceptibility of crops to infection by molds also favor fungal growth consequently mycotoxin production. Wounding of crops like tree nuts by birds, mammals, insects, may eventuate significant fungal infections. On the other hand, some insects carry mycotoxin (like aflatoxin) producing fungi associated with mycotoxin contamination in crops. The environmental factors affecting this contamination such as variation of seasons, disposal close to farm regions and the rates of insect population are all influenced by climate [22, 23]. Climate condition may directly influence some agricultural product by developing structural changes in crop. In this regard the hull cracking in nuts such as pistachio so called "early split" favor the fungal growth species especially *Aspergillus* spp. The rainfall that occurs at harvesting time may accelerate the fungal growth of crops. However in some geographical regions the time of high raining and high relative humidity of weather should be considered at harvesting and storage time of crops.

6.6 Storage

To avoid further mycotoxin contamination the agricultural product should be dried or de hulled immediately. The moisture content of stored food products must be lower than critical moisture (15% moisture and preferably to <13%) content according to water activity need for fungal growth (generally less than 0.7 at 25°C). In this situation the competition of microorganism for water not only prevents further growth of fungi but also inhibits the mycotoxin production [24]. Therefore in storage stage the main preventive action must be to decrease the moisture content and also the temperature. In Storage, any migration of moisture, condensation of moisture, and leaks should not occur.

The production of aflatoxin is strongly influenced by water activity of food commodities at storage time. On the other hand the infected food commodities specially crops and also nuts are able to provide adequate inoculums for incidence of the fungus to sound ones during poor storage practices [25]. Storage management is essential in preventing fungal proliferation and mycotoxin formation in any harvested products.

6.7 Transportation

For many foods may face mycotoxin problems, the transportation conditions and time are great factors controlling the increase of mycotoxins content. During the transportation, some extrinsic factors like moisture content, relative humidity, temperature and hygienic control (cleanliness, insect control, etc.) could directly affect the safety of food. It is much emphasized that the transportation of foods under high humidity may result in mycotoxins increase. Transportation must be done in controlled conditions and any failure in this part may lead to decay of high volume of commodities.

6.8 Processing

Since mycotoxins are chemical and thermal resistant, they can be stable during heat, physical and chemical processing of food so, the prevention of mycotoxin production in row food is a critical control point in food production chain. In terms of food safety it should be considered that most treatment of foods such as roasting, boiling, pasteurization, irradiation, freezing, drying, blanching, exhausting, boiling, curing, foaming, frying are not effective on elimination nor reduction of mycotoxin in contaminated food [3, 15, 19]. In this regard, it is better that all reduction or preventive strategies be performed before processing of food. On the other hand, these fungal toxins can also enter the human body via contaminated animal products (e.g., meat, egg, milk) due to feeding with mycotoxin contaminated feeds.

6.9 Prevention of mycotoxins

According to numerous reports about high occurrence of mycotoxins in foods/feeds they are a constant concern worldwide. Although the mycotoxin producer molds spores are present all over the environment and related toxins can be formed on crops during harvest stages, Storage, processing. Also the mold spores are present in soil and plant debris able to infect growing agriculture products simply and fast at any point of handling. However, mycotoxin contaminations cover most of economic costs including the practices of prevention and mitigation, the reduced volume of contaminated foods, animal feed contamination and reduction in animal performance or health effects. Nowadays management of mycotoxins involves all actions of prevention, regulation, control, monitoring, tracing, avoidance, decontamination, detoxification and animal treatments. Even at such total management there may be levels of mycotoxin in food products unavoidably as a continual concern [23].

The most efficient tool for mycotoxin problems is the prevention of mold growth in fields especially during postharvest practices. Additionally, environmental factors can immensely affect the production of mycotoxin by fungal species. In terms of predictive proceeding the predictive models have been developed as decision supporting systems to plan proper crop protection strategies in fields [25]. Innovative detection and diagnostic tools are also available to monitor the occurrence of mycotoxigenic fungi in fields and after harvest. When contamination is not prevented, several approaches can be employed to help remove mycotoxins from the contaminated commodities, including physical, chemical, and biological techniques. Detoxification processes should destroy or inactivate mycotoxins, by guaranteeing the nutritional value of food. Research is needed to study the fate of mycotoxins during decontamination, detoxification, and food processing. A holistic approach should be adopted to monitor, prevent, and control mycotoxigenic fungi and mycotoxins in food products.

Acknowledgements

The authors greatly thank Semnan University of Medical Sciences.

Conflict of interest

There is no conflict of interest.

Author details

Anna Abdolshahi* and Behdad Shokrollahi Yancheshmeh
Food Safety Research Center (Salt), Semnan University of Medical Sciences, Semnan, Iran

*Address all correspondence to: ana.abdoshahi@gmail.com

References

[1] Safety evaluation of certain food additives and contaminants. WHO Food Additives Series (on INCHEM.org). 1998;**40**:897-913

[2] Panel EB. Scientific opinion on an update on the present knowledge on the occurrence and control of foodborne viruses. EFSA Journal. 2011;**9**(7):2190-2196

[3] Jones JM. Food Safety. St. Paul. MN, USA: Eagan Press; 1992

[4] Montville TJ, Matthews KR. Food Microbiology: An Introduction. ASM Press; 2007

[5] Concon JM. Food Toxicology. Part A: Principles and Concepts; Part B: Contaminants and Additives. USA: Marcel Dekker Inc; 1988

[6] Gracias KS, McKillip JL. A review of conventional detection and enumeration methods for pathogenic bacteria in food. Canadian Journal of Microbiology. 2004;**50**(11):883-890

[7] Multari RA et al. Detection of biological contaminants on foods and food surfaces using laser-induced breakdown spectroscopy (LIBS). Journal of Agricultural and Food Chemistry. 2013;**61**(36):8687-8694

[8] Scallan E et al. Foodborne illness acquired in the United States—Major pathogens. Emerging Infectious Diseases. 2011;**17**(1):7

[9] Newell DG et al. Food-borne diseases—The challenges of 20 years ago still persist while new ones continue to emerge. International Journal of Food Microbiology. 2010;**139**:S3-S15

[10] Koopmans M, Duizer E. Foodborne viruses: An emerging problem. International Journal of Food Microbiology. 2004;**90**(1):23-41

[11] Marcogliese DJ, Pietrock M. Combined effects of parasites and contaminants on animal health: Parasites do matter. Trends in Parasitology. 2011;**27**(3):123-130

[12] Di Stefano V, Avellone G. Food contaminants. Journal of Food Studies. 2014;**3**:88-102

[13] Bánáti D. Consumer response to food scandals and scares. Trends in Food Science & Technology. 2011;**22**(2-3):56-60

[14] Seltzer JM. Biological contaminants. Journal of Allergy and Clinical Immunology. 1994;**94**(2):318-326

[15] Marin S et al. Mycotoxins: Occurrence, toxicology, and exposure assessment. Food and Chemical Toxicology. 2013;**60**:218-237

[16] Zain ME. Impact of mycotoxins on humans and animals. Journal of Saudi Chemical Society. 2011;**15**(2): 129-144

[17] Abdolshahi A et al. Aflatoxin binding efficiency of *Saccharomyces cerevisiae* mannoprotein in contaminated pistachio nuts. Food Control. 2018;**87**:17-21

[18] Abdolshahi A et al. Antifungal activities of coating incorporated with *Saccharomyces cerevisiae* cell wall mannoprotein on *Aspergillus flavus* growth and aflatoxin production in pistachio (*Pistacia vera* L.). Journal of Food Safety. 2019;**39**(2)e12608

[19] Rychlik M. Mycotoxins except Fusarium toxins in foods. In: Chemical Contaminants and Residues in Food. United Kingdom: Elsevier, Woodhead Publishing; 2017. pp. 279-294

[20] Schatzmayr G, Streit E. Global occurrence of mycotoxins in the food

and feed chain: Facts and figures. World Mycotoxin Journal. 2013;**6**(3):213-222

[21] Milićević DR, Škrinjar M, Baltić T. Real and perceived risks for mycotoxin contamination in foods and feeds: Challenges for food safety control. Toxins. 2010;**2**(4):572-592

[22] Galvano F, Ritieni A, Piva G, Pietri A. Mycotoxins in the human food chain. In: Diaz D. editor. Mycotoxin Blue Book. UK: Nottingham University Press; 2005. pp. 187-224

[23] Whitlow LW. Evaluation of mycotoxin binders. In: Proceedings of the 4th Mid-Atlantic Nutrition Conference; 2006

[24] Aldars-García L et al. An attempt to model the probability of growth and aflatoxin B1 production of *Aspergillus flavus* under non-isothermal conditions in pistachio nuts. Food Microbiology. 2015;**51**:117-129

[25] Aldars-García L et al. Modelling the probability of growth and aflatoxin B1 production of *Aspergillus flavus* under changing temperature conditions in pistachio nuts. Procedia Food Science. 2016;7:76-79

Chapter 3

Food Safety: The Risk of Mycotoxin Contamination in Fish

Constanze Pietsch

Abstract

Mycotoxins are commonly found in animal feeds, and fish feeds are no exception to this. The need to feed fish in aquaculture with compounded feeds leads to the increasing inclusion of plant-derived feed ingredients that have a higher probability of containing mycotoxins. Since fish appear to be quite sensitive to mycotoxins, further research on mycotoxin toxicity in fish is recommended. Depending on the chemical characteristics of an individual mycotoxin and the biotransformation abilities of the different fish species, certain mycotoxins can could be found in the edible parts of a fish. Thus, the consumption of fish products increases the potential risk of mycotoxin exposure for humans. This chapter reviews the risks associated with different groups of mycotoxins and makes recommendations on how to minimize these risks in the future.

Keywords: fish, aquaculture, mycotoxin toxicity, toxin residues

1. Introduction

Estimating risk requires sufficient knowledge of the frequency with which mycotoxins occur and the levels that can be expected. However, sufficiently detailed information on the actual levels of contamination in fish feeds is often not available. In addition, there is a high degree of variability between mycotoxins due to differences in fungal distribution and climatic conditions worldwide. Nevertheless, the following sections will summarize our current knowledge of mycotoxin occurrence in feed ingredients, fish feeds, and fish tissues in order to compile sufficient evidence to prove that some mycotoxins pose a considerable risk for consumers due to their high prevalence, incidence, toxicity, and/or stability as they pass into the food chain.

2. Exposure of fish to mycotoxins

Fish production in aquaculture has increased rapidly over the previous decades. Consequently, increasing numbers of fish have to be fed in aquaculture, which requires an increasing amount of fish feed. Since the global availability of fishmeal, which is a major ingredient in fish feed, is limited, cereals are common alternatives. Based on recent estimations, it has been determined that fishmeal is still a major component in fish feed in Europe [1], despite the fact that its percentage in commercial feeds has decreased over the last decades. The disadvantage of plant-based ingredients is that there is a higher probability of them being contaminated with mycotoxins. The second most prominent feed ingredient in aquaculture feeds in

Europe is wheat flour [1], followed by soybean products. Other feed ingredients are often present in fish feeds at average percentages of less than 10%, and these ingredients may also contain considerable amounts of mycotoxins. One example of such a problematic feed ingredient may be distillers' grain with solubles (DDGS) [1, 2].

The most important mycotoxins in feed ingredients in terms of risk to fish and consumers, since they are either known to be toxic and/or occur at high concentrations, include aflatoxin B1 (AFB_1), deoxynivalenol (DON), nivalenol (NIV), zearalenone (ZEN), ochratoxin A (OTA), T-2 toxin (T2), fumonisin B_1 (FB_1), moniliformin (MON), enniatins (ENNs), and beauvericin (BEA). Nevertheless, there are a number of reasons why mycotoxin contamination levels in feed ingredients can vary widely, for example, different fungal species or strains often grow on specific feed ingredients. Especially, high OTA levels have been found in corn (up to 1850 μg/kg, [3]), followed by wheat (up to 1024 μg/kg, [4]), soybean, and sunflower products (up to 350 and 240 μg/kg, respectively, [3]). Furthermore, *Fusarium* mycotoxins can contaminate peas and soybeans [5], and FB_1 can be found in significant amounts in corn [6].

The occurrence of mycotoxins in feed ingredients is also known to vary as a result of climate effects and differences in the distribution of various fungal species and strains that have differing abilities to form toxins [7–9]. The problem with mycotoxin contamination in feed ingredients is thought to have increased as a result of climate changes and the shipping of commodities on a global scale, which has led to the worldwide distribution of many fungal species, often resulting in higher contamination in cereals [9–11]. However, the presence of mycotoxins in feed ingredients does not mean that these substances will also be present in compounded animal feeds, since a number of mycotoxins have been reported to possess different degrees of stability when thermally processed and extruded [12]. Furthermore, the processing of feed ingredients, which includes cleaning, sorting, milling, and the application of thermal processes, also influences the mycotoxin load in the final products [13–16]. Nevertheless, the extent of the reduction in mycotoxin contamination during these procedures differs widely for each mycotoxin [15, 17–20]. Generally, mycotoxins that are most stable and widely distributed and, in most cases, occur at high concentrations in certain feed ingredients are problematic for fish production. Two mycotoxins that are already problematic at relatively low concentrations in fish feeds and will be reviewed in the section on fish toxicity are AFB_1 and OTA due to their high toxicity.

The most prominent member of the fumonisins in naturally contaminated animal feeds is FB_1 [21], which often occurs at high concentrations in feed ingredients (e.g., [22, 23]). However, since fumonisins are relatively unstable and easily affected by feed production processes, they are assumed to be less problematic than other mycotoxins. Nonetheless, feed processing may yield mycotoxin metabolites, in some cases resulting in increased toxicity [24].

ZEN is a mycotoxin that commonly occurs after crops have been infected have been infected with *Fusarium* species in the field, but this toxin can also develop during the storage of the cereals [25, 26]. ZEN contamination appears to be common in commercial fish feeds [27, 28], which raises concerns about the effects of chronic exposure to this mycotoxin, since besides exhibiting toxic characteristics, it is also a potent natural estrogen [29].

The trichothecenes include some very important mycotoxins, such as T-2 toxin, DON, and NIV. Recent research has focused on DON since it is known for its high prevalence and incidence in feed ingredients and animal feeds in Europe [30]. However, *Fusarium* fungi are also known to produce some less commonly described mycotoxins, known as emerging mycotoxins, which include BEA, ENNs, and MON [31, 32]. Although ENNs and BEA have been reported to be extremely prevalent in cereals [33], there has not been enough detailed research into their presence in feed components, compounded animal feeds, or farmed animals that have been exposed to these mycotoxins. The other important *Fusarium*-related mycotoxin is MON. Up to 1.2 mg/kg MON has been detected in feeds for higher vertebrates [34], whereas the levels present in commercial fish feeds remain unknown.

As mentioned above, mycotoxin contamination often occurs on crop fields, but improper storage of feed ingredients and feeds also contributes to the final toxin levels in fish diets. Toxin production depends on the fungi's ability to produce certain chemical compounds as well as environmental factors, such as physical, chemical, and biological factors [35]. Accordingly, similar to the aflatoxins, the occurrence of OTA seems to be connected to temperature and humidity in the environment during growth and harvesting of crops, and the storage of feed ingredients and feeds. However, for most investigated fish feeds, low OTA levels have been observed [28]. In contrast, recent research has shown that inappropriate storage over a period of 6 weeks of a commercial feed for salmonids can lead to the development of considerable amounts of OTA (up to 400 μg/kg feed, unpublished results, C. Pietsch).

Although dietary contamination is the main route of exposure for fish in aquaculture, mycotoxins may also be introduced to aquatic environments directly. For example, levels of 90 μg/L OTA have been reported in waste water originating from wine production. Furthermore, ZEN can be found in surface waters and in waste-water treatment plants at ng/L levels, which may be environmentally relevant due to the estrogenic effects of this mycotoxin [36–38]. Thus, the stability of mycotoxins in water may also have an effect on relevant exposure concentrations in aquatic environments [39].

When data on contamination levels and incidence in common feed ingredients are compiled, there may be significant uncertainties due to the fact that these studies use different methodologies for mycotoxin detection and quantification. Another problem when compiling data from scientific studies is that several studies have not reported accuracy and reliability parameters for their methods, meaning the measured toxin values probably contain uncertainties, since the sample preparation and detection procedures differed. Furthermore, actual mycotoxin concentrations in feed components, animal feeds, and animal tissues are often underestimated, since matrix effects and the problems of detecting masked mycotoxins, which can often not be detected by routine measurement techniques. Since research is continuously improving detection methods for mycotoxins, an increased number of comparative studies addressing the advantages and disadvantages of detection methods for more commonly and emerging mycotoxins, such as can be found in the study by Pascale [40], should be conducted.

Another problem with estimating actual contamination levels in feeds and animal tissues is that metabolites of even commonly occurring mycotoxins are often not analyzed together with their parent compound, although metabolites may occur in significant amounts as has been shown for DON [41]. Furthermore, toxin levels in the control diets used in experimental fish studies have often been reported to contain no mycotoxins, despite the fact that the necessary toxin analyses were rarely performed to provide proof for this assumption. This may lead to an underestimation of the actual toxin levels in both control diets and experimental diets if only a restricted number of mycotoxins are measured. As a result, actual mycotoxin exposure data for fish contain various uncertainties. Therefore, more complete feed contamination databases are required so that risk assessments can be improved.

3. Presence of mycotoxins and their toxicity in fish

If the risk to humans by consuming fish products is to be calculated, the first step would be to estimate the uptake and retention of mycotoxins in different fish

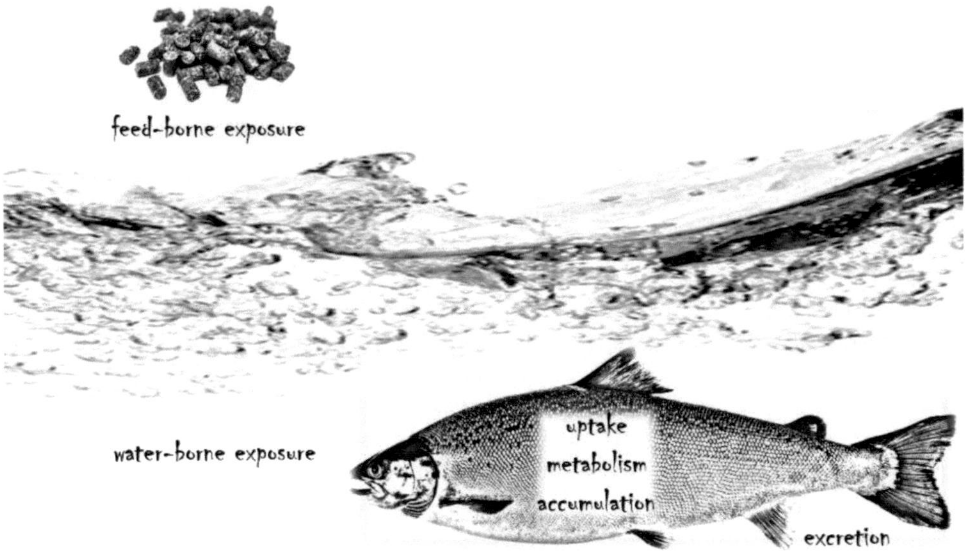

Figure 1.
Exposure routes and factors influencing mycotoxin retention in fish.

species and in different parts of the fish (**Figure 1**). Therefore, the following sections will summarize what is known about chemical characteristics in fish bodies and the toxicity in the animals resulting from the most important mycotoxins.

DON has a mean lowest-observable effect level (LOEL) in fish of 3541 ± 776 µg/kg (±SEM; **Figure 2**), whereas the contamination levels in commercial fish feeds range from 0 to 825 µg/kg [27, 28, 41]. Similar to findings in chickens, DON appears to be excreted rapidly by carp (*Cyprinus carpio*), leaving no relevant residues in the edible parts [42, 43]. FB_1 metabolization also occurs quickly in chicken and the remaining values in tissues stay low. However, exact information on the kinetics or biotransformation of fumonisins in fish is not available [44, 45]. Due to this and the large differences in the toxicity of fumonisins in fish (**Figure 2**), no exact risk can be calculated for farmed fish [1]. Typical disorders in higher vertebrates resulting from FB_1 exposure have often been linked to the disruption of the sphingolipid metabolism [46], and similar effects have also been observed in fish [47]. Nevertheless, a low potential risk has been assumed for most vertebrates, with the exception of pigs [45]. Despite the fact that the guidance values for fumonisins in complete fish feeds have been set by the European Commission and the US to 10 mg/kg based, some countries have chosen to set different guidance levels [48, 49]. Although FB_1 can affect fish at low concentrations, for example in carp (exposed to 500 µg/kg [50, 51]), the concentration range of the lowest-observable effects in fish is relatively broad, with a mean range of 26,480 ± 7124 µg/kg (±SEM; **Figure 2**), a level that is not achieved for either actual or estimated natural contamination of fish feeds [1, 52].

Previous studies have reported lethal concentrations of OTA that lead to 50% mortality (LC_{50}) ranging from 2 to 58 mg/kg body weight in various higher vertebrate species [53, 54]. Fish species appear to be particularly sensitive to OTA, and since disposition appears to mainly take place in the kidneys of fish and not in muscles [55], this not only affects its toxicity, but is also relevant for food safety. High sensitivity to OTA in fish has been demonstrated in several studies. The LC_{50} value for OTA in adult seabass (*Dicentrarchus labrax* L.) was found to be 280 µg/kg body weight [56], 360 µg/l for zebrafish (*Danio rerio*) embryos [57], and 5.53 mg/kg

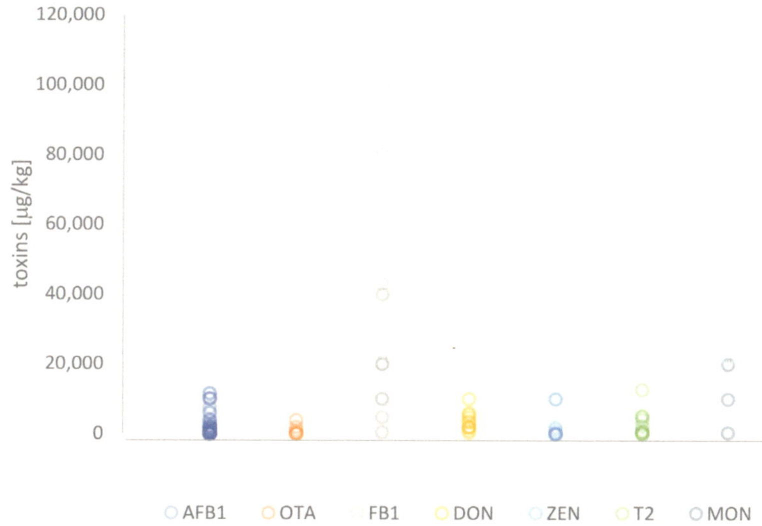

Figure 2.
Variability in mycotoxin toxicity for fish, as shown by the differences in the lowest-observable effect levels (LOEL) in different fish species. References: 92 studies for AFB$_1$ [63, 64, 70–149] comprising 21 different fish species, 7 studies for OTA [56–58, 94, 150–152] comprising 5 fish species, 15 studies for FB$_1$ [47, 50, 51, 153–165] reporting levels for 7 fish species, 12 studies for DON [42, 144, 166–175] yielding information for 5 different fish species, 10 studies for ZEN [144, 176–184] reporting LOEL for 5 different species, 10 studies [185–193] reporting effects of different levels of T-2 toxins on 4 different fish species, and 3 studies [162, 194, 195] for 3 different species exposed to MON.

body weight in rainbow trout (*Oncorhynchus mykiss*) [58]. However, the route of exposure may play a role when comparing these different studies. Furthermore, the absorption efficiency in the gut also determines the bioavailability of the mycotoxins in fish, as has been demonstrated for oral exposure to OTA in common carp [59]. If the LOEL for exposure of fish to OTA are summarized (**Figure 2**), the mean range is 1077 ± 566 µg/kg (±SEM), which indicates that the currently recommended guidance value for OTA in cereals and cereal products intended for animal feed of 250 µg/kg does not protect fish from potential damage [48]. This is in stark contrast to the guidance level of 20 µg/kg that exists in some non-EU countries [49].

ZEN has a mean toxicity value of 2389 ± 1285 µg/kg (±SEM), based on the LOEL calculations for five different fish species shown in **Figure 2**. Although the number of studies reporting effects of ZEN in fish is very limited, they may indicate that fish are more sensitive to water-borne ZEN than to dietary ZEN, which is why the mean LOEL level, including both, dietary and water-borne exposure for fish, shows quite a high standard error of the mean. ZEN concentrations above the LOEL levels in water samples have not been reported for aquatic environments [36–38]. Although the actual ZEN contamination of commercial fish feeds appears not to exceed the current guidance level for this mycotoxin in cereals and cereal products in the EU of 2000 µg/kg [27, 48], dietary exposure to this mycotoxin may still do harm to farmed fish. The guidance values in other countries that recommend maximum ZEN levels of 20–1000 µg/kg have a higher probability of protecting fish from damage [49], since the ZEN levels in fish feeds often do not exceed concentrations of 200 µg/kg [27, 60]. Nevertheless, more exact reports on ZEN toxicity in fish and the actual contamination levels in commercial fish feeds are needed to support these assumptions.

T-2 toxin has a mean toxicity of 3201 ± 1236 µg/kg (±SEM) in fish, based on the currently available LOEL for different fish species (**Figure 2**). This level is considerably higher than the actual contamination level found in salmonid fish feed in South America [28], and much lower than the guidance levels of 250 mg/kg for T-2 toxin set by the European Commission for cereal products in compound feeds [61] and individual recommendations in other countries (max. 80–100 mg/kg) for T-2 toxin in complete feed and all grains [49]. From these data, it can be assumed that fish do not regularly suffer from T-2 toxicity, and there have been no reports of accumulation of this mycotoxin in edible parts of the fish.

The situation for AFB_1 is, however, quite different. The mean LOEL for fish has been calculated to be 1248 ± 275 μg/kg (±SEM) (**Figure 2**). However, AFB_1 appears to be readily absorbed by the intestine [62] and a LOEL of less than 1 μg/kg has been observed in Nile tilapia (*Oreochromis niloticus*) and rainbow trout [63, 64], which shows that this mycotoxin can be a problem for farmed fish. In commercial fish feeds, AFB_1 levels are commonly less than 10 μg/kg [65, 66], but may be considerably higher in some cases [67–69]. Critical levels for fish have been estimated to be a mean of 4.30 μg/kg in commercial feeds [1], which indicates that farmed fish are exposed to a risk from AFB_1 intoxication.

Less information is available on the toxicity of ENNs and BEA in fish, but from initial experiments it can be assumed that at least some ENN toxins have toxic effects on zebrafish embryos (unpublished results, C. Pietsch). However, how relevant this toxicity is in comparison to the actual ENN contamination in commercial feeds remains unclear. Similar to other emerging mycotoxins, these substances have already been detected in the plasma of pigs after exposure to ENNs [196], indicating that the uptake of these substances occurs in vertebrates. In addition, it has been shown that food processing affects the presence on ENNs and BEA in bread [197, 198], and thermal processes, in particular, also appear to influence the ENN content in fish tissue [199]. Finally, the presence of high ENN and BEA levels in feed ingredients appears to overestimate the actual risk of fish feed contamination and the potential effects on farmed fish [1]. Thus, more research is needed on the toxicology and the biotransformation of ENNs and BEA in vertebrates.

An issue that also makes mycotoxin research difficult is the fact that we do not know enough about mycotoxin mixtures and their effects. Natural contamination of feed ingredients leads to the occurrence of several mycotoxins at the same time and their interactions remain mostly unknown.

4. Fish products and food safety

Exposure assessments are often based on a deterministic approach, which obtains the estimated daily intake (EDI) levels by assuming a human body weight of 60 kg for an adult. The EDI of each mycotoxin is commonly calculated as μg/kg body weight per day for each mycotoxin. Accordingly, the Joint FAO/WHO Expert Committee and Food Additives and Scientific Committee on Food have established a tolerable weekly intake (TWI) levels for humans for OTA of 120 ng/kg body weight and tolerable daily intake (TDI) levels of 250 ng/kg body weight for ZEN, 100 ng/kg body weight for T-2 and HT-2 toxins together, and 1000 ng/kg body weight for DON [200, 201]. For aflatoxins, no tolerable intake levels have been set since these toxins are listed as human carcinogens. The tolerable intake levels should be compared to the actual contamination levels found in fish products. However, the frequency of mycotoxin occurrence in fish products has not been investigated in detail. Recent studies indicate that less than 10% of fish and meat food samples are contaminated with mycotoxins, with DON contamination occurring in 17% of the 29 fish samples [202]. In addition, the accuracy of the reports also strongly depends on the accuracy and the number of samples that were analyzed.

Even if fish are exposed to feed-borne mycotoxins, and the resulting effects are not great, possible retention of these toxins in edible parts of the fish may pose a risk for human consumption. A risk to humans is assumed when the toxin concentrations in food exceed the safety limits. For AFB_1, this level has been set at 2 μg/kg by the European Union for food designated for human consumption [49]. However, the exact risk to humans is difficult to predict, since the behavior of the chemicals in the fish strongly depends on the chemical structures of the mycotoxins. In addition, toxin concentration in the feeds and duration of exposure also play an important role, therefore different studies may lead to different results. One example is the absence of accumulation of aflatoxin in the musculature of common carp in the study by Svobodova and Piskac [136], which contradicts the findings of Akter et al. [91]. The AFB_1 content in the hepatopancreas of gibel carp (*Carassius auratus gibelio*) was found to be considerably higher than in their muscle tissues (2.4–11.8 μg/kg) after 12 weeks of oral exposure [104]. An extrahepatic deposition of AFB_1 has also been confirmed in trout [62, 203], but the detection of this toxin in kidneys is more relevant from a toxicological point of view than from a food safety point of view. The study by Selim et al. [121] showed that exposure to 200 μg/kg AFB_1 for 2 weeks was sufficient to lead to detectable toxin residues in fish musculature (>20 μg/kg AFB_1), which increased to levels of more than 90 μg/kg AFB_1 after 10 weeks of exposure. Furthermore, feeding European seabass (*Dicentrarchus labrax* L.) with 18 μg/kg body weight AFB_1 resulted in toxin concentrations of 2.5 μg/kg AFB_1 in the fish musculature after 28 days of feeding, and even higher levels of 4.25 μg/kg AFB_1 after 42 days of exposure [94]. Compared to this, oral exposure of lambari fish (*Astyanax altiparanae*) to AFB_1 increased the body residues after feeding for at least 90 days [204]. In addition, this study showed that feeding an AFB_1 concentration of 50 μg/kg feed for 120 days also resulted in aflatoxin accumulation in muscle and liver tissues that were as high as in the feed. In other fish species, residues exceeding the safety limit were detected in the liver but not in the fish musculature [89, 104]. From these studies, it can be concluded that aflatoxin contamination can be a threat to humans after fish have been fed AFB_1 contaminated diets for certain duration. These values show that consuming fish can considerably add to the toxicological burden that can already be expected from consuming cereals, for which the daily intake through consumption of cereal-based products has been reported to reach levels of up to 7.9 ng/kg body weight [205] and 3 ng/kg body weight if peanuts are consumed [206]. An interesting finding was described in a study using walleye (*Sander vitreus*) which had been exposed to considerable amounts of AFB_1 that had accumulated in their edible parts. The accumulation of AFB_1 in the musculature may be reversible by feeding mycotoxin-free diets for 2 weeks [107], which also confirms similar findings in other fish species [104].

Fish muscle did not contain OTA in a Polish study [207]. In seabass (*Dicentrarchus labrax*) and sea bream (*Sparus aurata*) muscles, only low OTA levels have been detected [208]. It has already been reported that contaminated cereals and feed ingredients lead to the introduction of OTA into the food chain, posing a risk for humans [209]. Consuming fish appears to contribute to the presence of OTA in the food chain and also adds to the detectable levels of OTA in humans [2]. However, compared to the daily intake through direct consumption of cereal-based products that has been reported to be up to 22.2 ng/kg body weight for OTA [205], the amount that fish products may contribute to the toxicological burden appears to be lower. Nevertheless, this adds to the earlier assumption that naturally contaminated feeds also lead to the introduction of this mycotoxin into the food chain which may pose a risk to human consumers [210, 211]. The knowledge presented here on the presence and toxicity of this toxin in fish supports this assumption. The potential risk due to OTA exposure is probably caused by the fact that OTA is even more stable in the environment than aflatoxins [212, 213].

In contrast, the presence of fumonisins in fish appears not to be relevant for consumers, since they rarely occur in farmed fish (e.g., in a survey in Switzerland in only one fillet sample containing less than 0.06 μg/kg FB_1 + FB_2, personal communication C. Pietsch). In addition, it was not possible to identify a high risk to humans as a result of consuming fish products contaminated with other mycotoxins, such as ZEN and DON, since no relevant toxin levels could be detected in the musculature of DON- or ZEN-treated rainbow trout and common carp [42, 214, 215]. Interestingly, ZEN exposure did result in retention in the ovaries of farmed trout [184]. Furthermore, the study by Nácher-Mestre et al. [216] found no detectable mycotoxin levels in gilthead sea bream or Atlantic salmon (*Salmo salar*) after 8 months of dietary exposure to DON levels of up to 79.2 μg/kg and fumonisins at levels of up to 754 μg/kg. A study into fish as food reported mean DON levels of 1.19 μg/kg [202]; and since DON was the major mycotoxin in the fish samples analyzed in this study, it was also assumed to be the main contributor to the daily human mycotoxin exposure. ZEN retention in human breast milk has already been related to consuming meat, fish, dry fruits, and spices [217]. However, compared to the presence of *Fusarium* toxins in cereals, it can still be assumed, based on the fact that rapid metabolization takes place in fish, that the retention of DON and ZEN in fish is low. Therefore, there can be no assumption of a higher risk to humans of consuming these mycotoxins in fish compared to the risk of exceeding the toxicological reference values by consuming cereal products directly [202, 206, 218].

In the 29 fish samples in the study by Carballo et al. [202], mean ENN A concentrations of 0.89 μg/kg were observed. ENNs were also detected in 20% of the salmon flesh samples and 10% of rainbow trout samples in the study by Tolosa et al. [199], but further processing including cooking or smoking appears to mitigate the toxin content [219]. In contrast, fish from Egypt contained predominant xerophilic molds with *Aspergillus* species being the major ones (58.2%), followed by *Penicillium* species (32.7%) in salted products and also in smoke-cured bonga shad and African catfish (*Ethmalosa fimbriata* and *Clarias gariepinus*) [220, 221]. However, a study in Kenya only showed aflatoxins in dried fish, and not in fresh ones [222]. Smoked-dried fish from Nigeria may also contain potential mycotoxin producing fungi and aflatoxins [223–226]. Similar results from Egyptian smoked fish confirmed that the moisture and salt concentrations that occur during food processing influence the OTA and AFB_1 contents in the fish products, possibly exceeding the permissible limits for both mycotoxins [227].

Mycotoxins can also occur in sun-dried fish products, which are typically found in tropical and subtropical regions where high temperatures and humidity considerably influence fungal growth and toxin formation. Accordingly, samples of dried seafood contained high levels of ZEN and OTA (317.3 and 1.9 μg/kg, respectively). Furthermore, low amounts of AFB_2 (1.2 μg/kg) were also observed in the muscle of crucian carp (*Carassius carassius*), even after storage for 3 months at room temperature [228], emphasizing the high stability of aflatoxins.

5. Conclusions

Taken together, mycotoxin contamination in feed ingredients and fish feeds is an increasing problem that will have to be addressed by crop farmers, feed producers, and researchers. One step that could be taken is to prevent heavily contaminated raw materials being introduced into the feed production processes, which would lower potential mycotoxin contamination levels. Nevertheless, other mycotoxins are still formed during storage, and improved guidelines and recommendations for storage of feed ingredients and animal feeds should be published. Since mycotoxins are present in animal feeds, in some cases at toxicological relevant levels, this may cause health problems in fish and limit production in aquaculture. More data on the presence of mycotoxins in fish would allow better risk assessments for human consumers to be carried out. Furthermore, the data sets for some mycotoxins indicate that more strict guidance levels are needed for fish feeds to protect farm animals from harm and prevent accumulation of potentially problematic mycotoxins such as AFB_1 and OTA in the food chain.

Acknowledgements

Darren Mace's (ZHAW, Wädenswil, Switzerland) work on checking the language in the entire manuscript is highly appreciated.

Conflict of interest

The author declares that there are no conflicts of interest regarding the publication of this chapter.

Author details

Constanze Pietsch
Zurich University of Applied Sciences (ZHAW), Institute of Natural Resource Sciences (IUNR), Wädenswil, Switzerland

*Address all correspondence to: constanze.pietsch@zhaw.ch

References

[1] Pietsch C. Risk assessment for mycotoxin contamination in fish feeds in Europe. Mycotoxin Research. 2019. pp. 1-22. DOI: 10.1007/s12550-019-00368-6

[2] Mortensen A, Granby K, Eriksen FD, Cederberg TL, Friis-Wandall S, Simonsen Y, et al. Levels and risk assessment of chemical contaminants in byproducts for animal feed in enmark. Journal of Environmental Science and Health—Part B Pesticides, Food Contaminants, and Agricultural Wastes. 2014;**49**(11):797-810. DOI: 10.1080/03601234.2014.938546

[3] Rafai P, Bata Á, Jakab L, Ványi A. Evaluation of mycotoxin-contaminated cereals for their use in animal feeds in Hungary. Food Additives and Contaminants. 2000;**17**:799-808

[4] Czerwiecki L, Czajkowska D, Witkowska-Gwiazdowska A. On ochratoxin A and fungal flora in Polish cereals from conventional and ecological farms. Part 2: Occurrence of ochratoxin A and fungi in cereals in 1998. Food Additives and Contaminants. 2002;**19**:1051-1057

[5] Ivić D, Domijan AM, Peraica M, Miličević T, Cvjetković B. *Fusarium* spp. contamination of wheat, maize, soybean, and pea grain in Croatia. Arhiv za Higijenu Rada i Toksikologiju. 2009;**60**:435-442

[6] Nelson PE, Desjardins AE, Plattner RD. Fumonisins, mycotoxins produced by *Fusarium* species: Biology, chemistry, and significance. Annual Review of Phytopathology. 2003;**31**:233-252

[7] Schatzmayr G, Streit E. Global occurrence of mycotoxins in the food and feed chain: Facts and figures. World Mycotoxin Journal. 2013;**6**(3):213-222

[8] Anater A, Manyes L, Meca G, Ferrer E, Luciano FB, Pimpão CT, et al. Mycotoxins and their consequences in aquaculture: A review. Aquaculture. 2016;**451**:1-10

[9] Levic J, Gosic-Dondo S, Ivanovic D, Stankovic S, Krnjaja V, Bocarov-Stancic A, et al. An outbreak of *Aspergillus* species in response to environmental conditions in Serbia. Pestic i Fitomedicina. 2013;**28**(3):167-179

[10] Dobolyi C, Sebők F, Varga J, Kocsubé S, Szigeti G, Baranyi N, et al. Occurrence of aflatoxin producing *Aspergillus flavus* isolates in maize kernel in Hungary. Acta Alimentaria. 2013;**42**(3):451-459

[11] Tóth B, Török O, Kótai É, Varga M, Toldiné Tóth É, Pálfi X, et al. Role of Aspergilli and Penicillia in mycotoxin contamination of maize in Hungary. Acta Agron Hungarica. 2012;**60**(2):143-149

[12] Bullerman LB, Bianchini A. Stability of mycotoxins during food processing. International Journal of Food Microbiology. 2007;**119**(1-2):140-146

[13] Kaushik G. Effect of processing on mycotoxin content in grains. Critical Reviews in Food Science and Nutrition. 2015;**55**:1672-1683

[14] Kushiro M. Effects of milling and cooking processes on the deoxynivalenol content in wheat. International Journal of Molecular Sciences. 2008;**9**:2127-2145

[15] Cheli F, Pinotti L, Rossi L, Dell'Orto V. Effect of milling procedures on mycotoxin distribution in wheat fractions: A review. LWT—Food Science and Technology. 2013;**54**:307-314

[16] Mmongoyo JA, Wu F, Linz JE, Nair MG, Mugula JK, Tempelman RJ,

et al. Aflatoxin levels in sunflower seeds and cakes collected from micro- and small-scale sunflower oil processors in Tanzania. PLoS One. 2017;**12**(4):e0175801

[17] Lancova K, Hajslova J, Kostelanska M, Kohoutkova J, Nedelnik J, Moravcova H, et al. Fate of trichothecene mycotoxins during the processing: Milling and baking. Food Additives & Contaminants. Part A, Chemistry, Analysis, Control, Exposure & Risk Assessment. 2008;**25**:650-659

[18] Neuhof T, Koch M, Rasenko T, Nehls I. Occurrence of zearalenone in wheat kernels infected with Fusarium culmorum. World Mycotoxin Journal. 2008;**1**:429-435

[19] Edwards SG, Dickin ET, MacDonald S, Buttler D, Hazel CM, Patel S, et al. Distribution of Fusarium mycotoxins in UK wheat mill fractions. Food Additives & Contaminants. Part A, Chemistry, Analysis, Control, Exposure & Risk Assessment. 2011;**28**:1694-1704

[20] Pascale M, Haidukowski M, Lattanzio VMT, Silvestri M, Ranieri R, Visconti A. Distribution of T-2 and HT-2 toxins in milling fractions of durum wheat. Journal of Food Protection. 2011;**74**:1700-1707

[21] Griessler K, Encarnacao P. Fumonisins—Mycotoxins of increasing importance in fish. Aquaculture Asia Magazine. 2009;**XIV**:24-26

[22] Grajewski J, Blajet-Kosicka A, Twaruzek M, Kosicki R. Occurrence of mycotoxins in Polish animal feed in years 2006-2009. Journal of Animal Physiology and Animal Nutrition. 2012;**96**(5):870-877

[23] Stanković S, Lević J, Ivanović D, Krnjaja V, Stanković G, Tančić S. Fumonisin B_1 and its co-occurrence with other fusariotoxins in naturally-contaminated wheat grain. Food Control. 2012;**23**(2):384-388

[24] Scott P. Recent research on fumonisins: A review. Food Additives & Contaminants. Part A. 2012;**29**(2):242-248

[25] Caldwell RW, Tuite J, Stob M, Baldwin R. Zearalenone production by *Fusarium* species. Applied Microbiology. 1970;**20**(1):31-34

[26] Milano GD, Lopez TA. Influence of temperature on zearalenone production by regional strains of *Fusarium graminearum* and *Fusarium oxysporum* in culture. International Journal of Food Microbiology. 1991;**13**(4):329-333

[27] Pietsch C, Kersten S, Burkhardt-Holm P, Valenta H, Dänicke S. Occurrence of deoxynivalenol and zearalenone in commercial fish feed: An initial study. Toxins (Basel). 2013;**5**:184-192

[28] Greco M, Pardo A, Pose G. Mycotoxigenic fungi and natural co-occurrence of mycotoxins in rainbow trout (*Oncorhynchus mykiss*) feeds. Toxins (Basel). 2015;**7**:4595-4609. DOI: 10.3390/toxins7114595

[29] Gupta RC, Mostrom MS, Evans TJ. Chapter 76—Zearalenone. In: Gupta RC, editor. Veterinary Toxicology. Basic and Clinical Principles. 3rd Edition. Section XV Mycotoxins. Academic Press; 2018. pp. 1055-1063. DOI: 10.1016/B978-0-12-811410-0.00076-3

[30] Rodrigues I, Naehrer K. A three-year survey on the worldwide occurrence of mycotoxins in feedstuffs and feed. Toxins (Basel). 2012;**4**(9):663-675

[31] Polido OM, Gill S. Chapter 35—Food and toxicologic pathology: An overview. In: Haschek WM, Rousseaux CG, Wallig MA, editors. Haschek and Rousseaux's Handbook of Toxicologic Pathology. 3rd ed. Elsevier

Inc. Academic Press; 2013. pp. 1051-1076. DOI: 10.1016/C2010-1-67850-9

[32] Escrivá L, Font G, Manyes L. In vivo toxicity studies of fusarium mycotoxins in the last decade: A review. Food and Chemical Toxicology. 2015;**78**:185-206

[33] Lindblad M, Gidlund A, Sulyok M, Börjesson T, Krska R, Olsen M, et al. Deoxynivalenol and other selected fusarium toxins in swedish wheat—Occurrence and correlation to specific fusarium species. International Journal of Food Microbiology. 2013;**167**(2):284-291. DOI: 10.1016/j.ijfoodmicro.2013.07.002

[34] Labuda R, Parich A, Vekiru E, Tančinová D. Incidence of fumonisins, moniliformin and Fusarium species in poultry feed mixtures from Slovakia. Annals of Agricultural and Environmental Medicine. 2005;**12**(1):81-86

[35] Piotrowska M, Slizewska K, Biernasiak J. Mycotoxins in Cereal and Soybean-Based Food and Feed, Soybean - Pest Resistance. In: El-Shemy HA editor. IntechOpen; February 13th 2013. DOI: 10.5772/54470. Available from: https://www.intechopen.com/books/soybean-pest-resistance/mycotoxins-in-cereal-and-soybean-based-food-and-feed

[36] Bucheli TD, Erbs M, Hartmann N, Vogelgsang S, Wettstein FE, Forrer H-R. Estrogenic mycotoxins in the environment. Mitteilungen aus Lebensmitteluntersuchung und Hygiene. 2005;**96**:386-403

[37] Gromadzka K, Waśkiewicz A, Świetlik J, Bocianowski J, Goliński P. The role of wastewater treatment in reducing pollution of surface waters with zearalenone. Arhiv za Higijenu Rada i Toksikologiju. 2015;**66**(2):159-164. DOI: 10.1515/aiht-2015-66-2606

[38] Spengler P, Körner W, Metzger J. Substances with estrogenic activity in effluents of sewage treatment plants in southwestern Germany. 1. Chemical analysis. Environmental Toxicology and Chemistry. 2001;**20**:2133-2141

[39] Nogueira R, Estevinho I, Abrunhosa L, Mendonça C, Machado P, Carballa M, et al. Assessing the degradation of ochratoxin a using a bioassay: The case of contaminated winery wastewater. Water Science and Technology. 2007;**56**(2):55-61

[40] Pascale M. Detection methods for mycotoxins in cereal grains and cereal products. Zbornik Matice srpske za prirodne nauke. 2009;**117**:15-25. DOI: 10.2298/ZMSPN0917015P

[41] Mirocha CJ, Abbas HK, Windels CE, Xie W. Variation in deoxynivalenol, 15-acetyldeoxynivalenol, 3-acetyldeoxynivalenol, and zearalenone production by Fusarium graminearum isolates. Applied and Environmental Microbiology. 1989;**55**(5):1315-1316

[42] Pietsch C, Michel C, Kersten S, Valenta H, Dänicke S, Schulz C, et al. In vivo effects of deoxynivalenol (DON) on innate immune responses of carp (*Cyprinus carpio* L.). Food and Chemical Toxicology. 2014;**68**:44-52

[43] Awad WA, Ghareeb K, Böhm J, Razzazi E, Hellweg P, Zentek J. The impact of the Fusarium toxin deoxynivalenol (DON) on poultry. International Journal of Poultry Science. 2008;**7**(9):827-842

[44] Vudathala DK, Prelusky DB, Ayroud M, Trenholm HL, Miller JD. Pharmacokinetic fate and pathological effects of ^{14}C-fumonisin B1 in laying hens. Natural Toxins. 1994;**2**(2):81-88

[45] CONTAM P. (EFSA Panel on Contaminants in the Food Chain). Risks for animal health related to the presence of fumonisins, their modified forms and hidden forms in feed. EFSA Journal. 2018;**16**(5):5242

[46] Wang E, Ross PF, Wilson TM, Riley RT, Merrill AH. Increases in serum sphingosine and sphinganine and decreases in complex sphingolipids in ponies given feed containing fumonisins, mycotoxins produced by Fusarium moniliforme. The Journal of Nutrition. 1992;**122**:1706-1716

[47] Goel S, Lenz SD, Lumlertdacha S, Lovell RT, Shelby RA, Li M, et al. Sphingolipid levels in catfish consuming Fusarium moniliforme corn culture material containing fumonisins. Aquatic Toxicology. 1994;**30**(4):285-294

[48] European Commission. Commission Recommendation 2006/576/EC of 17 August 2006 on the presence of deoxynivalenol, zearalenone, ochratoxin A, T-2 and HT-2 and fumonisins in products intended for animal feeding. Official Journal of the European Union. 2006:L229/7-L229/9

[49] Van Egmond H, Jonker M. Worldwide Regulations for Mycotoxins in Food and Feed in 2003. Vol. 81. Rome, Italy: FAO Food and Nutrition Paper; 2004

[50] Pepeljnjak S, Petrinec Z, Kovacic S, Segvic M. Screening toxicity study in young carp (*Cyprinus carpio* L.) on feed amended with fumonisin B_1. Mycopathologia. 2003;**156**(2):139-145

[51] Abu-Hassan FMA, Khalil RH, Saad TT, Amer M, Abdel-Latif R. Histopathological outcomes designating the toxicological aspects of Fumonisin B_1 on cultured Nile tilapia, *Oreochromis niloticus*. International Journal of Fisheries and Aquatic Studies. 2016;**4**(3):52-60

[52] Rodríguez-Cervantes C, Girón-Pérez M, Robledo-Marenco M, Marín S, Velázquez-Fernández J, Medina-Díaz I, et al. Aflatoxin B 1 and its toxic effects on immune response of teleost fishes: A review. World Mycotoxin Journal. 2010;**3**(2):193-199

[53] Kuiper-Goodman T, Scott PM. Risk assessment of the mycotoxin ochratoxin A. Biomedical and Environmental Sciences. 1989;**2**:179-248

[54] Peckham JC, Doupnik B, Jones OH. Acute toxicity of ochratoxins A and B in chicks. Applied Microbiology. 1971;**21**:492-494

[55] Fuchs R, Appelgren L-E, Hult K. Distribution of 14C-ochratoxin A in the rainbow trout (Salmo gairdneri). Acta Pharmacologica et Toxicologica. 1986;**59**:220-221

[56] El-Sayed YS, Khalil RH, Saad TT. Acute toxicity of ochratoxin-A in marine water-reared sea bass (*Dicentrarchus labrax* L.). Chemosphere. 2009;**75**(7):878-882

[57] Tschirren L, Siebenmann S, Pietsch C. Toxicity of ochratoxin to early life stages of zebrafish (*Danio rerio*). Toxins (Basel). 2018;**10**(7):264

[58] Doster RC, Sinnhuber RO, Pawlowski NE. Acute intraperitoneal toxicity of ochratoxin a and B derivatives in rainbow trout (*Salmo gairdneri*). Food and Cosmetics Toxicology. 1974;**12**:499-505

[59] Hagelberg S, Hult K, Fuchs R. Toxicokinetics of ochratoxin A in several species and its plasma-binding properties. Journal of Applied Toxicology. 1989;**9**(2):91-96

[60] Kokic B, Cabarkapa I, Levic J, Mandic A, Matic J, Ivanov D. Screening of mycotoxins in animal feed from the region of Vojvodina. Matica Srpska Proceedings for Natural Sciences. 2010;**117**:87-96

[61] European Commission. Commission Recomendation No 2013/165/EU of March 2013 on the presence of T-2 and HT-2 toxin in cereals and cereal products. Official Journal of the European Union. 2013;**91**:12-15

[62] Ngethe S, Horsberg TE, Ingebrigtsen K. The disposition of ^{3}H-aflatoxin B1 in the rainbow trout (*Oncorhynchus mykiss*) after oral and intravenous administration. Aquaculture. 1992;**108**(3-4):323-332

[63] El-Enbaawy M, Adel M, Marzouk M, AA S. The effect of acute and chronic aflatoxicosis on the immune functions of Oreochromis niloticus in Egypt. Veterinary Medical Journal-Giza. 1994;**42**:47-52

[64] Imani A, Salimi Bani M, Noori F, Farzaneh M, Moghanlou KS. The effect of bentonite and yeast cell wall along with cinnamon oil on aflatoxicosis in rainbow trout (*Oncorhynchus mykiss*): Digestive enzymes, growth indices, nutritional performance and proximate body composition. Aquaculture. 2017;**476**:160-167

[65] María E, Gonçalves-Nunes C, Gomes-Pereira MM, Raposo-Costa AP, Da Rocha-Rosa CA, Pereyra CM, et al. Screening of aflatoxin B1 and mycobiota related to raw materials and finished feed destined for fish. Latin American Journal of Aquatic Research. 2015;**43**(3):595-600. DOI: 10.3856/vol43-issue3-fulltext-22

[66] Samuel TO, Odunigba O. Aflatoxins associated with storage fungi in fish feeds. IFE Journal of Science. 2015;**17**(2):519-523

[67] Altug G, Beklevik G. Level of aflatoxin in some fish feeds from fish farming processes, feed factories and imported feeds. Turkish Journal of Veterinary and Animal Sciences. 2003;**27**:1247-1252

[68] Gonçalves RA. Aflatoxins: A threat to yellow catfish production. World Aquaculture Society. 2016;**47**(1):56-57

[69] Marijani E, Wainaina JM, Charo-Karisa H, Nzayisenga L, Munguti J, Joselin Benoit Gnonlonfin G, et al. Mycoflora and mycotoxins in finished fish feed and feed ingredients from smallholder farms in East Africa. Egyptian Journal of Aquatic Research. 2017;**43**:169-176

[70] Ayyat MS, Abd Rahman GA, El-Marakby HI, El-Hakem NAB, Hessan AAA. Toxicity and biochemical hazards induced by exposure of Nile tilapia to aflatoxin and their amelioration. Global Journal of Environmental Sciences and Toxicology. 2014;**1**:1-19

[71] Ayyat MS, Ayyat AMN, Al-Sagheer AA, El-Hais AEAM. Effect of some safe feed additives on growth performance, blood biochemistry, and bioaccumulation of aflatoxin residues of Nile tilapia fed aflatoxin-B_1 contaminated diet. Aquaculture. 2018;**495**:27-34

[72] Baglodi V, EG J, KM N, PB A. Effect of dietary incorporated aflatoxin (AFB1) on the survival and growth performance of *Labeo rohita*. Journal of Experimental Zoology, India. 2015;**18**(2):603-607

[73] Bailey GS, Goeger DE, Hendricks JD. Factors influencing experimental carcinogenesis in laboratory fish models. In: Varanasi U, editor. Metabolism of Polycyclic Aromatic Hydrocarbons in the Aquatic Environment. Boca Raton: CRC Press; 1989. p. 253

[74] Bailey GS, Price RL, Park DL, Hendricks JD. Effect of ammoniation of aflatoxin B1-contaminated cottonseed feedstock on the aflatoxin M_1 content of cows' milk and hepatocarcinogenicity in the trout bioassay. Food and Chemical Toxicology. 1994;**32**(8):707-715

[75] Bailey GS, Dashwood R, Loveland MP, Pereira C, Hendricks DJ. Molecular dosimetry in fish: Quantitative target organ DNA

adduction and hepatocarcinogenicity for four aflatoxins by two exposure routes in rainbow trout. Mutation Research/Fundamental and Molecular Mechanisms of Mutagenesis. 1998;**399**(2):233-244

[76] Bailey GS, Williams DE, Hendricks JD. Fish models for environmental carcinogenesis: The rainbow trout. Environmental Health Perspectives. 1996;**104**(Suppl 1):5-21

[77] Bastardo H, Sofia S, Nava L, Rangel C. Effects of aflatoxin B_1 concentration and exposure time on hepatic condition in the rainbow trout. Interciencia. 2006;**31**(6):437-440

[78] Bauer DH, Lee DJ, Sinnhuber RO. Acute toxicity of aflatoxins B_1 and G_1 in the rainbow trout (*Salmo gairdneri*). Toxicology and Applied Pharmacology. 1969;**15**(2):415-419

[79] Bedoya-Sernal CM, Michelin E, Massocco M, Carrion L, Godoy S, Lima C, et al. Effects of dietary aflatoxin B1 on accumulation and performance in matrinxã fish (Brycon cephalus). PLoS One. 2018;**13**(8):e0201812

[80] Abd-Allah GA, El-Fayoumi RI, Smith MJ, Heckmann RA, O'Neill KL. A comparative evaluation of aflatoxin B1 genotoxicity in fish models using the comet assay. Mutation Research—Genetic Toxicology and Environmental Mutagenesis. 1999;**446**(2):181-188

[81] Black JJ, Maccubbin AE, Schiffert M. A reliable, efficient, microinjection apparatus and methodology for the in vivo exposure of rainbow trout and salmon embryos to chemical carcinogens. Journal of the National Cancer Institute. 1985;**75**(6):1123-1128

[82] Breinholt V, Hendricks J, Pereira C, Arbogast D, Bailey G. Dietary chlorophyllin is a potent inhibitor of aflatoxin B_1 hepatocarcinogenesis in rainbow trout. Cancer Research. 1995;**55**(1):57-62

[83] Cagauan AG, Tayaban RH, Somga JR, Bartolome RM. Effect of aflatoxin-contaminated feeds in Nile tilapia (*Oreochromis niloticus* L.). In: Abstract of the 6th International Symposium on Tilapia in Aquaculture (ISTA 6) Section: Health Management and Diseases Manila, Philippines; 12-16 September 2014. pp. 172-178

[84] Carpenter HM, Zhang Q, El Zahr C, Selivonchick DP, Brock DE, Curtis LR. *In vitro* and *in vivo* temperature modulation of hepatic metabolism and DNA adduction of aflatoxin B1 in rainbow trout. Journal of Biochemical Toxicology. 1995;**10**(1):1-10

[85] Chávez-Sánchez MC, Martínez Palacios CA, Osorio Moreno I. Pathological effects of feeding young *Oreochromis niloticus* diets supplemented with different levels of aflatoxin B1. Aquaculture. 1994;**127**(1):49-60

[86] Curtis LR, Zhang Q, El-Zahr C, Carpenter HM, Miranda CL, Buhler DR, et al. Temperature-modulated incidence of aflatoxin B1-initiated liver cancer in rainbow trout. Toxicological Sciences. 1995;**25**(1):146-153. DOI: 10.1093/toxsci/25.1.146

[87] Ottinger CA, Kaattari SL. Long-term immune dysfunction in rainbow trout (*Oncorhynchus mykiss*) exposed as embryos to aflatoxin B_1. Fish & Shellfish Immunology. 2000;**10**(1):101-106

[88] Dashwood RH, Fong AT, Williams DE, Hendricks JD, Bailey GS. Promotion of aflatoxin BI carcinogenesis by the natural tumor modulator indole-3-carbinol: Influence of dose, duration, and intermittent exposure on indole-3-carbinol promotional potency. Cancer Research. 1991;**51**(9):2362-2365

[89] Deng SX, Tian LX, Liu FJ, Jin SJ, Liang GY, Yang HJ, et al. Toxic effects

and residue of aflatoxin B_1 in tilapia (*Oreochromis niloticus* × *O. aureus*) during long-term dietary exposure. Aquaculture. 2010;**307**(3-4):233-240

[90] El-Barbary MI. Detoxification and antioxidant effects of garlic and curcumin in *Oreochromis niloticus* injected with aflatoxin B1 with reference to gene expression of glutathione peroxidase (GPx) by RT-PCR. Fish Physiology and Biochemistry. 2016;**42**(2):617-629. DOI: 10.1007/s10695-015-0164-4

[91] Akter A, Rahman M, Hasan M. Effects of aflatoxin B 1 on growth and bioaccumulation in common carp fingerling in Bangladesh. Asia-Pacific Journal of Rural Development. 2010;**20**(2):1-13

[92] El-Banna R, Teleb H, Hadi MM, Fakhry FM. Performance and tissue residue of tilapias fed dietary aflatoxin. Veterinary Medical Journal Giza. 1992;**40**(3):17-23

[93] El-Boshy M, El-Ashram A, Abd El-Ghany N. Effect of dietary beta1,3 glucan on immunomodulation od diseased *Oreochromins niloticus* experimentally infected with aflatoxin B_1. In: 8th International Symposium on Tilapia in Aquaculture. Cairo, Egypt; 2008. pp. 1109-1127

[94] El-Sayed YS, Khalil RH. Toxicity, biochemical effects and residue of aflatoxin B_1 in marine water-reared sea bass (*Dicentrarchus labrax* L.). Food and Chemical Toxicology. 2009;**47**:1606-1609

[95] Thorgaard GH, Arbogast DN, Hendricks JD, Pereira CB, Bailey GS. Tumor suppression in triploid trout. Aquatic Toxicology. 1999;**46**(2):121-126

[96] Goeger DE, Shelton DW, Hendricks JD, Pereira C, Bailey GS. Comparative effect of dietary butylated hydroxyanisole and β-naphthoflavone on aflatoxin b1 metabolism, DNA adduct formation, and carcinogenesis in rainbow trout. Carcinogenesis. 1988;**9**(10):1793-1800

[97] Gonçalves RA, Do Cam T, Tri NN, Santos GA, Encarnação P, Hung LT. Aflatoxin B1 (AFB_1) reduces growth performance, physiological response, and disease resistance in Tra catfish (*Pangasius hypophthalmus*). Aquaculture International. 2018;**26**(3):921-936

[98] Han D, Xie S, Zhu X, Yang Y, Guo Z. Growth and hepatopancreas performances of gibel carp fed diets containing low levels of aflatoxin B_1. Aquaculture Nutrition. 2010;**16**(4):335-342

[99] Hatanaka J, Doke N, Harada T, Aikawa T, Enomoto M. Usefulness and rapidity of screening for the toxicity and carcinogenicity of chemicals in medaka *Oryzias latipes*. The Japanese Journal of Experimental Medicine. 1982;**52**:243-253

[100] Hegazi S, El-Sabagh M, El-Keeidy A, Zein El-Dein A. Aflatoxin in feed and its effect on fish health. Kafr El-Sheikh Veterinary Medical Journal. 2013;**11**(2):317-329

[101] Hendricks JD, Wales JH, Sinnhuber RO, Nixon JE, Loveland PM, Scanlan RA. Rainbow trout (*Salmo gairdneri*) embryos: A sensitive animal model for experimental carcinogenesis. Federation Proceedings. 1980;**39**(14):3222-3229

[102] Farabi SMV, Yousefian M, Hajimoradloo A. Aflatoxicosis in juvenile *Huso huso* fed a contaminated diet. Journal of Applied Ichthyology. 2006;**22**(S1):234-237

[103] Hendricks JD, Arbogast DN, Pereira CB, Bailey GS. Long-term, high-dose dietary exposure of rainbow trout to butylated hydroxyanisole in non-carcinogenic. Cancer Letters. 1994;**78**(1-3, 193):189

[104] Huang Y, Han D, Zhu X, Yang Y, Jin J, Chen Y, et al. Response and recovery of gibel carp from subchronic oral administration of aflatoxin B_1. Aquaculture. 2011;**319**(1-2):89-97

[105] Hussain D, Mateen A. Alleviation of aflatoxin-B_1 toxicity by using clay adsorbent in nile tilapia (*Oreochromis niloticus*) diets. Pakistan Journal of Zoology. 2017;**49**(2):425-431

[106] Hussein S, Mekkawy I, Moktar Z, Mubarak M. Protective effect of *Nigella sativa* seed against aflatoxicosis in *Oreochromis niloticus*. In: Mycotoxin Conference Mycotoxins & Environ. Poland Bydgoszez. 2000. Vol. 25-27. pp. 109-130

[107] Hussain M, Gabal MA, Wilson T, Summerfelt RC. Effect of aflatoxin-contaminated feed on morbidity and residues in walleye fish. Veterinary and Human Toxicology. 1993;**35**(5):396-398

[108] Jantrarotai W, Lovell RT, Grizzle JM. Acute toxicity of aflatoxin B_1 to channel catfish. Journal of Aquatic Animal Health. 1990;**2**(4):237-247

[109] Jantrarotai W, Lovell RT. Subchronic toxicity of dietary aflatoxin B_1 to channel catfish. Journal of Aquatic Animal Health. 1990;**2**(4):248-254

[110] Lim HA, Ng WK, Lim SL, Ibrahim CO. Contamination of palm kernel meal with *Aspergillus flavus* affects its nutritive value in pelleted feed for tilapia, *Oreochromis mossambicus*. Aquaculture Research. 2001;**32**(11):895-905

[111] Madhusudhanan N, KavithaLakshmi SN, Shanmugasundaram KR, Shanmugasundaram E. Oxidative damage to lipids and proteins induced by aflatoxin B-1 in fish (*Labeo rohita*)-protective role of Amrita Bindu. Environmental Toxicology and Pharmacology. 2004;**17**(2):73-77

[112] Mahfouz ME. Ameliorative effect of curcumin on aflatoxin B_1-induced changes in liver gene expression of *Oreochromis niloticus*. Molecular Biology. 2015;**49**(2):275-286

[113] Al Faragi JK. The efficacy of prebiotic (β-glucan) as a feed additive against toxicity of aflatoxin B1 in common carp, *Cyprinus carpio* L. Journal of Aquaculture Research and Development. 2014;**5**(4):1000240-1000246

[114] Mahfouz ME, Sherif AH. A multiparameter investigation into adverse effects of aflatoxin on *Oreochromis niloticus* health status. Journal of Basic and Applied Zoology. 2015;**71**:48-59

[115] Magouz F, Eweedah N, Salem M, AA A. Detoxification of aflatoxin contaminated ration by chemical, biological and spices methods in Nile tilapia (*Oreochromis niloticus*) diets. Journal of Agricultural Research. Kafr El-Shaikh University. 2016;**42**(4):102-119

[116] Marijani E, Nasimolo J, Kigadye E, Gnonlonfin GJB, Okoth S. Sex-related differences in hematological parameters and organosomatic indices of *Oreochromis niloticus* exposed to aflatoxin B_1 diet. Scientifica (Cairo). 2017 Article ID 4268926, 7 pages

[117] Mohapatra S, Sahu NP, Pal AK, Prusty AK, Kumar V, Kumar S. Haemato-immunology and histo-architectural changes in Labeo rohita fingerlings: Effect of dietary aflatoxin and mould inhibitor. Fish Physiology and Biochemistry. 2011;**37**(1):177-186

[118] Shelton DW, Goeger DE, Hendricks JD, Bailey GS. Mechanisms of anti-carcinogenesis: The distribution and metabolism of aflatoxin B1 in rainbow trout fed aroclor 1254. Carcinogenesis. 1986;**7**(7):1065-1071

[119] Shahafve S, Banaee M, Haghi B, Mohiseni M. Histopathological study of common carp (*Cyprinus carpio*) fed aflatoxin-contaminated diets. International Journal of Aquatic Biology. 2017;**5**(2):63-70

[120] Sepahdari A, Ebrahimzadeh Mosavi HA, Sharifpour I, Khosravi A, Motallebi AA, Mohseni M, et al. Effects of different dietary levels of AFB1 on survival rate and growth factors of Beluga (*Huso huso*). Iranian Journal of Fisheries Sciences. 2010;**9**(1):141-150

[121] Selim KM, El-hofy H, Khalil RH. The efficacy of three mycotoxin adsorbents to alleviate aflatoxin B_1-induced toxicity in *Oreochromis niloticus*. Aquaculture International. 2014;**22**(2):523-540

[122] Schoenhard G, Hendricks J, Nixon J, Lee D, Wales J, Sinnhuber R, et al. Aflatoxicol-induced hepatocellular carcinoma in rainbow trout (*Salmo gairdneri*) and the synergistic effects of cyclopropenoid fatty acids. Cancer Research. 1981;**3**:1011-1014

[123] Sato S, Matsushima T, Tanaka N, Sugimura T, Takashima F. Hepatic tumors in the guppy (*Lebistes reticulatus*) induced by aflatoxin B_1, dimethylnitrosamine, and 2-acetylaminofluorene. Journal of the National Cancer Institute. 1973;**50**(3):767-778. DOI: 10.1093/jnci/50.3.767

[124] Andleeb S, Ashraf M, Hafeez-Ur-Rehman M, Jabbar MA, Abbas F, Younus M. Effect of aflatoxin B_1-contaminated feed on growth and vital organs of advance fry of *Catla catla*. Journal of Animal and Plant Sciences. 2015;**25**(3):816-825

[125] Sahoo P, Mukherjee S, Jain A, Mukherjee A. Histopathological and electron microscopic studies of gills and opisthonephros of rohu, Labeo rohita to acute and subchronic aflatoxin B_1 toxicity. Asian Fisheries Science. 2003;**16**:257-268

[126] Sahoo PK, Mukherjee SC. Immunosuppressive effects of aflatoxin B_1 in Indian major carp (*Labeo rohita*). Comparative Immunology, Microbiology and Infectious Diseases. 2001;**24**(3):143-149

[127] Sahoo PK, Mukherjee SC. The effect of dietary immunomodulation upon Edwardsiella tarda vaccination in healthy and immunocompromised Indian major carp (*Labeo rohita*). Fish & Shellfish Immunology. 2002;**12**(1):1-16

[128] Sahoo PK, Mukherjee SC. Immunomodulation by dietary vitamin C in healthy and aflatoxin B_1-induced immunocompromised rohu (*Labeo rohita*). Comparative Immunology, Microbiology and Infectious Diseases. 2003;**26**(1):65-76

[129] Royes J-AB, Yanong RPE. Molds in fish feeds and aflatoxicosis. In: Fact Sheet FA-95 [Internet]. 2002. Available from: http://edis.ifas.ufl.edu

[130] Shelton D, Hendricks J, Coulombe R, Bailey G. Effect of dose on the inhibition of carcinogenesis/mutagenesis by Aroclor 1254 in rainbow trout fed aflatoxin B_1. Journal of Toxicology and Environmental Health. 1984;**13**:649-657

[131] Sinnhuber R, Wales J, Ayres J, Engebrecht R, DL A. Dietary factors and hepatoma in rainbow trout (*Salmo gairdneri*). 1. Aflatoxins in vegetable protein feedstuffs. Journal of the National Cancer Institute. 1968;**41**:711-718

[132] Spring P, Fegan DF. Mycotoxins—A rising threat to aquaculture. Nutritional biotechnology in the feed and food industries. In: Proceedings of Alltech's 21st Annual Symposium, Lexington, Kentucky, USA, 22-25 May 2005. pp. 323-331

[133] Weigt S, Huebler N, Strecker R, Braunbeck T, Broschard TH. Zebrafish (*Danio rerio*) embryos as a model for testing proteratogens. Toxicology. 2011;**281**(1-3):25-36. DOI: 10.1016/j.tox.2011.01.004

[134] Svobodova Z, Piskac A, Havlikova J, Groch L. Influence of feed with different contents of B_1 aflatoxin on the carp health condition. Zivocisna Vyroba–UVTIZ. 1982;**27**(11):811-820

[135] Arana S, Tabata YA, Sabino M, Rigolino MG, Hernandez-Blazquez FJ. Differential effect of chronic aflatoxin B_1 intoxication on the growth performance and incidence of hepatic lesions in triploid and diploid rainbow trout (*Oncorhynchus mykiss*). Archivos de Medicina Veterinaria. 2002;**XXXIV**(2):253-263

[136] Svobodova Z, Piskac A. Effect of feeds with a low content of aflatoxin B1 on the health of carp *Cyprinus carpio*. Zivocisna Vyroba–UVTIZ. 1980;**25**(11):809-814

[137] Tilton SC, Gerwick LG, Hendricks JD, Rosato CS, Corley-Smith G, Givan SA, et al. Use of a rainbow trout oligonucleotide microarray to determine transcriptional patterns in aflatoxin B1-induced hepatocellular carcinoma compared to adjacent liver. Toxicological Sciences. 2005;**88**(2):319-330

[138] Troxel CM, Reddy AP, O'Neal PE, Hendricks JD, Bailey GS. In vivo aflatoxin B_1 metabolism and hepatic DNA adduction in zebrafish (*Danio rerio*). Toxicology and Applied Pharmacology. 1997;**143**(1):213-220

[139] Varior S, Philip B. Aflatoxin B_1 induced alterations in the stability of the lysosomal membrane in *Oreochromis mossambicus* (Peters, 1852). Aquaculture Research. 2012;**43**(8):1170-1175

[140] Wang X. Response of yellow catfish (*Pelteobagrus fulvidraco*) to different dietary concentrations of aflatoxin B1 and evaluation of an aflatoxin binder in offsetting its negative effects. Ciencias Marinas. 2016;**42**(1):15-29

[141] Zhang Q, Suorsa-Super K, Curtis LR. Temperature-modulated aflatoxin B1 hepatic disposition and formation and persistence of DNA adducts in rainbow trout. Toxicology and Applied Pharmacology. 1992;**113**(2):253-259

[142] Zychowski KE, Rodrigues Hoffmann A, Ly HJ, Pohlenz C, Buentello A, Romoser A, et al. The effect of aflatoxin-B1 on red drum (*Sciaenops ocellatus*) and assessment of dietary supplementation of NovaSil for the prevention of aflatoxicosis. Toxins (Basel). 2013;**5**(9):1555-1573. DOI: 10.3390/toxins5091555

[143] Zychowski KE, Pohlenz C, Mays T, Romoser A, Hume M, Buentello A, et al. The effect of NovaSil dietary supplementation on the growth and health performance of Nile tilapia (*Oreochromis niloticus*) fed aflatoxin-B_1 contaminated feed. Aquaculture. 2013;**376-379**:117-123

[144] Zhou H, George S, Li C, Gurusamy S, Sun X, Gong Z, et al. Combined toxicity of prevalent mycotoxins studied in fish cell line and zebrafish larvae revealed that type of interactions is dose-dependent. Aquatic Toxicology. 2017;**193**:60-71

[145] Lopes PRS, Pouey JLOF, Enke DBS, Mallmann CA, Kich HA, Soquetta MB. Utilização de adsorvente em rações contendo aflatoxina para alevinos de jundiá Use of adsorbent in diets containing aflatoxin for silver catfish fingerlings. Revista Brasileira de Zoologia. 2009;**38**(4):589-595

[146] Ayres JL, Lee DJ, Wales JH, Sinnhuber RO. Aflatoxin structure and hepatocarcinogenicity in rainbow trout (*Salmo gairdneri*). Journal

of the National Cancer Institute. 1971;**46**(3):561-564. DOI: 10.1093/jnci/46.3.561

[147] Kövesi B, Pelyhe C, Zándoki E, Mézes M, Balogh K. Changes of lipid peroxidation and glutathione redox system, and expression of glutathione peroxidase regulatory genes as effect of short-term aflatoxin B_1 exposure in common carp. Toxicon. 2018;**144**:103-108

[148] Sherif A, Abdel-Maksoud AS, Mustafa S. Study on toxicity of *Oreochromis niloticus* with aflatoxin B1. Egyptian Journal of Aquatic Biology and Fisheries. 2013;**17**:107-119. DOI: 10.12816/0011065

[149] Ayyat MS, Rhman GAA, El-Marakby HI, Hessein AAA. Aflatoxin B1 toxicity and its reduction by using coumarin and vitamin E in Nile tilapia. Zagazig Journal of Agricultural Research. 2017;**41**(1):20

[150] Haq M, Gonzalez N, Mintz K, Jaja-Chimedza A, De Jesus CL, Lydon C, et al. Teratogenicity of ochratoxin a and the degradation product, ochratoxin α, in the zebrafish (*Danio rerio*) embryo model of vertebrate development. Toxins (Basel). 2016;**8**(2):40. DOI: 10.3390/toxins8020040

[151] Manning BB, Ulloa RM, Li MH, Robinson EH, Rottinghaus GE. Ochratoxin A fed to channel catfish (*Ictalurus punctatus*) causes reduced growth and lesions of hepatopancreatic tissue. Aquaculture. 2003;**219**(1-4):739-750

[152] Zahran E, Manning B, Seo JK, Noga EJ. The effect of Ochratoxin A on antimicrobial polypeptide expression and resistance to water mold infection in channel catfish (*Ictalurus punctatus*). Fish & Shellfish Immunology. 2016;**57**:60-67

[153] Gbore FA, Adewole AM, Oginni O, Oguntolu MF, Bada AM, Akele O. Growth performance, haematology and serum biochemistry of African catfish (*Clarias gariepinus*) fingerlings fed graded levels of dietary fumonisin B 1. Mycotoxin Research. 2010;**26**(4):221-227

[154] Kovačić S, Pepeljnjak S, Petrinec Z, Klarić MŠ. Fumonisin B1 neurotoxicity in young carp (*Cyprinus carpio* L.). Archives of Industrial Hygiene and Toxicology. 2009;**60**(4):419-426

[155] Li MH, Raverty SA, Robinson EH. Effects of dietary mycotoxins produced by the mold *Fusarium moniliforme* on channel catfish *Ictalurus punctatus*. Journal of the World Aquaculture Society. 1994;**25**(4):512-516

[156] Yildirim M, Lim C, Wan PJ, Klesius PH. Growth performance and immune response of channel catfish (*Ictalurus puctatus*) fed diets containing graded levels of gossypol-acetic acid. Aquaculture. 2003;**219**(1-4):751-768

[157] Cirra Scaff RM, Scussel VM. Ultra-structural and histochemical analysis of channel catfish (*Ictalurus punctatus*) liver treated with fumonisin B1. Brazilian Archives of Biology and Technology. 2008;**51**(2):333-344. DOI: 10.1590/S1516-89132008000200013

[158] Claudino-Silva SC, Lala B, Mora NHAP, Schamber CR, Nascimento CS, Pereira VV, et al. Challenge with fumonisins B_1 and B_2 changes IGF-1 and GHR mRNA expression in liver of Nile tilapia fingerlings. World Mycotoxin Journal. 2018;**11**(2):237-245

[159] Adeyemo BT, Tiamiyu LO, Ayuba VO. Serum biochemistry and lipids profiling in experimental dietary exposure of *Heterobranchus longifilis* catfish juveniles to graded concentrations of fumonisin B_1. International Journal of Aquaculture. 2017;7(5):31-41. DOI: 10.5376/ija.2017.07.0005

[160] Carrera García E. Effects of fumonisin B1 on performance of juvenile Baltic salmon (*Salmo salar*) [thesis]. Finland: Master of Science in Sustainable Management of Inland Aquatic Resources; 2013. DOI: 10.13140/RG.2.1

[161] McKean C, Tang L, Tang M, Billam M, Wang Z, Theodorakis CW, et al. Comparative acute and combinative toxicity of aflatoxin B_1 and fumonisin B_1 in animals and human cells. Food and Chemical Toxicology. 2006;**44**:868-876

[162] Tuan NA, Manning BB, Lovell RT, Rottinghaus GE. Responses of Nile tilapia (*Oreochromis niloticus*) fed diets containing different concentrations of moniliformin or fumonisin B1. Aquaculture. 2003;**217**:515-528

[163] Petrinec Z, Pepeljnjak S, Kovacic S, Krznaric A. Fumonisin B_1 causes multiple lesions in common carp (*Cyprinus carpio*). Deutsche Tierärztliche Wochenschrift. 2004;**111**:358-363

[164] Lumlertdacha S, Lovell RT. Fumonisin-contaminated dietary corn reduced survival and antibody production by channel catfish challenged with *Edwardsiella ictaluri*. Journal of Aquatic Animal Health. 1995;**7**(1):1-8

[165] Lumlertdacha S, Lovell RT, Shelby RA, Lenz SD, Kemppainen BW. Growth, hematology, and histopathology of channel catfish, *Ictalurus punctatus*, fed toxins from *Fusarium moniliforme*. Aquaculture. 1995;**130**:201-218

[166] Bernhoft A, Høgåsen HR, Rosenlund G, Ivanova L, Berntssen MHG, Alexander J, et al. Tissue distribution and elimination of deoxynivalenol and ochratoxin A in dietary-exposed Atlantic salmon (*Salmo salar*). Food Additives & Contaminants: Part A: Chemistry, Analysis, Control, Exposure & Risk Assessment. 2017;**34**(7):1211-1224

[167] Woodward B, Young LG, Lun AK. Vomitoxin in diets for rainbow trout (*Salmo gairdneri*). Aquaculture. 1983;**35**:93-101

[168] Döll S, Valenta H, Baardsen G, Möller P, Koppe W, Stubhaug I, et al. Effects of increasing concentrations of deoxynivalenol, zearalenone and ochratoxin A in diets for Atlantic salmon (*Salmo salar*) on performance, health and toxin residues. In: Proceedings of 33rd Mycotoxin Workshop. Freising, Germany: Society for Mycotoxin Research; 2011

[169] Gonçalves RA, Navarro-Guillén C, Gilannejad N, Dias J, Schatzmayr D, Bichl G, et al. Impact of deoxynivalenol on rainbow trout: Growth performance, digestibility, key gene expression regulation and metabolism. Aquaculture. 2018;**490**:362-372

[170] Manning BB, Abbas HK, Wise DJ, Greenway T. The effect of feeding diets containing deoxynivalenol contaminated corn on channel catfish (*Ictalurus punctatus*) challenged with *Edwardsiella ictaluri*. Aquaculture Research. 2013;**45**(11):1782-1786

[171] Matejova I, Vicenova M, Vojtek L, Kudlackova H, Nedbalcova K, Faldyna M, et al. Effect of the mycotoxin deoxynivalenol on the immune responses of rainbow trout (*Oncorhynchus mykiss*). Veterinarni Medicina. 2015;**60**(9):515-521

[172] Ryerse IA, Hooft JM, Bureau DP, Hayes MA, Lumsden JS. Purified deoxynivalenol or feed restriction reduces mortality in rainbow trout, *Oncorhynchus mykiss* (Walbaum), with experimental bacterial Coldwater disease but biologically relevant concentrations of deoxynivalenol do not impair the growth of *Flavobacter*.

Journal of Fish Diseases. 2015;**38**: 809-819. DOI: 10.1111/jfd.12295

[173] Ryerse IA, Hooft JM, Bureau DP, Anthony Hayes M, Lumsden JS. Diets containing corn naturally contaminated with deoxynivalenol reduces the susceptibility of rainbow trout (*Oncorhynchus mykiss*) to experimental *Flavobacterium psychrophilum* infection. Aquaculture Research. 2016;**47**(3):787-796

[174] Sanden M, Jørgensen S, Hemre GI, Ørnsrud R, Sissener NH. Zebrafish (*Danio rerio*) as a model for investigating dietary toxic effects of deoxynivalenol contamination in aquaculture feeds. Food and Chemical Toxicology. 2012;**50**(12):4441-4448

[175] Šišperová E, Modrá H, Ziková A, Kloas W, Blahová J, Matejová I, et al. The effect of mycotoxin deoxynivalenol (DON) on the oxidative stress markers in rainbow trout (Oncorhynchus mykiss, Walbaum 1792). Journal of Applied Ichthyology. 2015;**31**(5):855-861

[176] Pietsch C, Junge R. Physiological responses of carp (*Cyprinus carpio* L.) to dietary exposure to zearalenone (ZEN). Comparative Biochemistry and Physiology—Part C: Toxicology & Pharmacology. 2016;**188**:52-59

[177] Pietsch C. Zearalenone (ZEN) and its influence on regulation of gene expression in carp (*Cyprinus carpio* L.) liver tissue. Toxins (Basel). 2017;**9**:283-300

[178] Arukwe A, Grotmol T, Haugen TB, Knudsen FR, Goksøyr A. Fish model for assessing the in vivo estrogenic potency of the mycotoxin zearalenone and its metabolites. Science of the Total Environment. 1999;**236**(1-3):153-161

[179] Bakos K, Kovács R, Staszny Á, Sipos DK, Urbányi B, Müller F, et al. Developmental toxicity and estrogenic potency of zearalenone in zebrafish (*Danio rerio*). Aquatic Toxicology. 2013;**136-137**:13-21

[180] Chen H, Hu J, Yang J, Wang Y, Xu H, Jiang Q, et al. Generation of a fluorescent transgenic zebrafish for detection of environmental estrogens. Aquatic Toxicology. 2010;**96**(1):53-61

[181] Johns SM, Denslow ND, Kane MD, Watanabe KH, Orlando EF, Sepúlveda MS. Effects of estrogens and antiestrogens on gene expression of fathead minnow (*Pimephales promelas*) early life stages. Environmental Toxicology. 2011;**26**(2):195-206

[182] Schwartz P, Thorpe KL, Bucheli TD, Wettstein FE, Burkhardt-Holm P. Short-term exposure to the environmentally relevant estrogenic mycotoxin zearalenone impairs reproduction in fish. Science of the Total Environment. 2010;**409**(2):326-333

[183] Woźny M, Brzuzan P, Gusiatin M, Jakimiuk E, Dobosz S, Kuźmiński H. Influence of zearalenone on selected biochemical parameters in juvenile rainbow trout (*Oncorhynchus mykiss*). Polish Journal of Veterinary Sciences. 2012;**15**(2):221-225

[184] Woźny M, Obremski K, Jakimiuk E, Gusiatin M, Brzuzan P. Zearalenone contamination in rainbow trout farms in North-Eastern Poland. Aquaculture. 2013;**416-417**:209-211

[185] Yuan G, Wang Y, Yuan X, Zhang T, Zhao J, Huang L, et al. T-2 toxin induces developmental toxicity and apoptosis in zebrafish embryos. Journal of Environmental Sciences (China). 2014;**26**(4):917-925

[186] Balogh K, Heincinger M, Fodor J, Mézes M. Effect of long term feeding of T-2 and HT-2 toxin contaminated diet on the glutathione redox status and lipid peroxidation processes in

common carp (*Cyprinus carpio* L.). Acta Biologica Szegediensis. 2009;**53**

[187] Pelyhe C, Kövesi B, Zándoki E, Kovács B, Szabó-Fodor J, Mézes M, et al. Effectof4-weekfeedingofdeoxynivalenol-or T-2-toxin-contaminated diet on lipid peroxidation and glutathione redox system in the hepatopancreas of common carp (*Cyprinus carpio* L.). Mycotoxin Research. 2016;**32**(2):77-83. DOI: 10.1007/s12550-016-0242-1

[188] Pelyhe C, Kövesi B, Zándoki E, Kovács B, Szabó-Fodor J, Mézes M, et al. Short-term effects of T-2 toxin or deoxynivalenol on lipid peroxidation and the glutathione system in common carp. Acta Veterinaria Hungarica. 2016;**64**(4):449-466

[189] Matejova I, Faldyna M, Modra H, Blahova J, Palikova M, Markova Z, et al. Effect of T-2 toxin-contaminated diet on common carp (*Cyprinus carpio* L.). Fish and Shellfish Immunology. 2017;**60**:458-465

[190] Kravchenko L, Galash V, Avreneva L, Kranauskas A. On the sensitivity of carp, *Cyprinus carpio*, to mycotoxin T-2. Journal of Ichthyology. 1989;**29**:156-160

[191] Modra H, Sisperova E, Blahova J, Enevova V, Fictum P, Franc A, et al. Elevated concentrations of T-2 toxin cause oxidative stress in the rainbow trout (*Oncorhynchus mykiss*). Aquaculture Nutrition. 2018;**24**(2):842-849

[192] Marasas WFO, Bamburg JR, Smalley EB, Strong FM, Ragland WL, Degurse PE. Toxic effects on trout, rats, and mice of T-2 toxin produced by the fungus *Fusarium tricinctum* (Cd.) Snyd. et Hans. Toxicology and Applied Pharmacology. 1969;**15**(2):471-482

[193] Manning BB, Li MH, Robinson EH, Gaunt PS, Camus AC, Rottinghaus GE. Response of channel catfish to diets containing T-2 toxin. Journal of Aquatic Animal Health. 2003;**15**(3):229-238

[194] Gonçalves R, Tarasco M, Schatzmayr D, Gavaia P. Preliminary evaluation of moniliformin as a potential threat for teleosts. Fishes. 2018;**3**(1):4. DOI: 10.3390/fishes3010004

[195] Yildirim M, Manning BB, Lovell RT, Grizzle JM, Rottinghaus GE. Toxicity of moniliformin and fumonisin B_1 fed singly and in combination in diets for young channel catfish *Ictalurus punctatus*. Journal of the World Aquaculture Society. 2010;**31**(4):599-608

[196] Devreese M, De Baere S, De Backer P, Croubels S. Quantitative determination of the Fusarium mycotoxins beauvericin, enniatin A, A_1, B and B_1 in pig plasma using high performance liquid chromatography-tandem mass spectrometry. Talanta. 2013;**106**:212-219. DOI: 10.1016/j.talanta.2012.11.068

[197] Meca G, Ritieni A, Mañes J. Influence of the heat treatment on the degradation of the minor Fusarium mycotoxin beauvericin. Food Control. 2012;**28**(1):13-18

[198] Hu L, Koehler P, Rychlik M. Effect of sourdough processing and baking on the content of enniatins and beauvericin in wheat and rye bread. European Food Research and Technology. 2014;**238**(4):581-587

[199] Tolosa J, Font G, Mañes J, Ferrer E. Mitigation of enniatins in edible fish tissues by thermal processes and identification of degradation products. Food and Chemical Toxicology. 2017;**101**:67-74

[200] Joint FAO/WHO Expert Committee on Food Additives (JECFA). Safety evaluation of certain mycotoxins in food. WHO Food Addit. 2001;Ser. 47

[201] European Committee Scientific Committee on Food. Opinion on Fusarium toxins. Part 6: Group evaluation of T-2 toxin, HT-2 toxin, nivalenol and deoxynivalenol. Eur Comm. 2002;1-12

[202] Carballo D, Moltó JC, Berrada H, Ferrer E. Presence of mycotoxins in ready-to-eat food and subsequent risk assessment. Food and Chemical Toxicology. 2018;**121**:558-565

[203] Larsson P, Ngethe S, Ingebrigtsen K, Tjälve H. Extrahepatic disposition of ^{3}H-aflatoxin B_1 in the rainbow trout (*Oncorhynchus mykiss*). Pharmacology & Toxicology. 1992;**71**(4):262-271

[204] Michelin EC, Massocco MM, Godoy SHS, Baldin JC, Yasui GS, Lima CG, et al. Carryover of aflatoxins from feed to Lambari fish (*Astyanax altiparanae*) tissues. Food Additives & Contaminants: Part A: Chemistry, Analysis, Control, Exposure & Risk Assessment. 2017;**34**(2):265-272

[205] Huong BTM, Tuyen LD, Tuan DH, Brimer L, Dalsgaard A. Dietary exposure to aflatoxin B_1, ochratoxin A and fuminisins of adults in Lao Cai province, Viet Nam: A total dietary study approach. Food and Chemical Toxicology. 2016;**98**(Pt B):127-133. DOI: 10.1016/j.fct.2016.10.012

[206] Caldas ED, Jardim ANO. Exposure to toxic chemicals in the diet: Is the brazilian population at risk. Journal of Exposure Science and Environmental Epidemiology. 2012;**22**(1):1-15. DOI: 10.1038/jes.2011.35

[207] Wiśniewska-Dmytrow H, Zmudzki J, Burek O, Pietruszka K. Official control of ochratoxin A in food of animal origin in Poland between 2003 and 2012. Bulletin of the Veterinary Institute in Pulawy. 2013;**57**:519-523

[208] Guardone L, Tinacci L, Luci G, Meucci V, Intorre L, Armani A. Determination of ochratoxin A in European seabass and gilthead seabream in Italy. In: Poster by the European College of Veterinary Public Health. 2018. DOI: 10.13140/RG.2.2.31611.77604

[209] Reddy L, Bhoola K. Ochratoxins-food contaminants: Impact on human health. Toxins. 2010;**2**(4):771-779

[210] Duarte SC, Lino CM, Pena A. Food safety implications of ochratoxin A in animal-derived food products. Veterinary Journal. 2012;**192**:286-292

[211] Duarte SC, Lino CM, Pena A. Ochratoxin A in feed of food-producing animals: An undesirable mycotoxin with health and performance effects. Veterinary Microbiology. 2011;**154**(1-2):1-13

[212] Duarte SC, Pena A, Lino CM. A review on ochratoxin A occurrence and effects of processing of cereal and cereal derived food products. Food Microbiology. 2010;**27**:187-198

[213] Moss MO. Risk assessment for aflatoxins in foodstuffs. International Biodeterioration and Biodegradation. 2002;**50**(3-4):137-142

[214] Pietsch C, Kersten S, Valenta H, Dänicke S, Schulz C, Burkhardt-Holm P, et al. Effects of dietary exposure to zearalenone (ZEN) on carp (*Cyprinus carpio* L.). Toxins (Basel). 2015;7(9):3465-3480

[215] Woźny M, Dobosz S, Obremski K, Hliwa P, Gomułka P, Łakomiak A, et al. Feed-borne exposure to zearalenone leads to advanced ovarian development and limited histopathological changes in the liver of premarket size rainbow trout. Aquaculture. 2015;**448**:71-81

[216] Nácher-Mestre J, Serrano R, Beltrán E, Pérez-Sánchez J, Silva J, Karalazos V, et al. Occurrence and potential transfer of mycotoxins in gilthead sea bream and Atlantic salmon by

use of novel alternative feed ingredients. Chemosphere. 2015;**128**:314-320

[217] Dinleyici M, Aydemir O, Yildirim GK, Kaya TB, Carman KB. Human mature milk zearalenone and deoxynivalenol levels in Turkey. Neuroendocrinology Letters. 2018;**39**(4):325-330

[218] Raad F, Nasreddine L, Hilan C, Bartosik M, Parent-Massin D. Dietary exposure to aflatoxins, ochratoxin A and deoxynivalenol from a total diet study in an adult urban lebanese population. Food and Chemical Toxicology. 2014;**73**:35-43

[219] Tolosa J, Barba FJ, Font G, Ferrer E. Mycotoxin incidence in some fish products: QuEChERS methodology and liquid chromatography linear ion trap tandem mass spectrometry approach. Molecules. 2019;**24**:527. DOI: 10.3390/molecules24030527

[220] Ahmed AM, Ismail SA, Abd-El-Rahman HAEL. Quantitative, qualitative and toxigenic evaluations of xerophilic mold in traditional Egyptian salted fish, Molouha. Journal of Food Safety. 2005;**25**(1):9-18

[221] Edema MO, Agbon AO. Significance of fungi associated with smoke-cured *Ethmalosa fimbriata* and *Clarias gariepinus*. Journal of Food Processing & Preservation. 2010;**34**(S1):355-363

[222] Orony DNA, Lalah JO, Jondiko IO. Determination of carcinogenic polycyclic aromatic hydrocarbons (PAHs), aflatoxins, and nitrosamines in processed fish from the Winam Gulf area of Kenya and estimated potential exposure in human. Polycyclic Aromatic Compounds. 2016;**36**:295-317

[223] Akinyemi AA, Adejola AQ, Obasa SO, Ezeri GNO. Aflatoxins in smoked-dried fish sold in Abeokuta, Ogun State, South-west Nigeria. COLERM Proceedings. 2012;**2**:478-487

[224] Olajuyigbe OO, Akande GR, Ezekiel CN, Ezekiel MO. Aflatoxigenic moulds and aflatoxin contamination of retailed fishery products in Lagos markets. Mycotoxicology Society of Nigeria Mycotoxicology. 2014;**1**:57-63

[225] Adebayo-Tayo BC, Onilude AA, Patrick UG. Mycofloral of smoke-dried fishes sold in Uyo, Eastern Nigeria. World Journal of Agricultural Sciences. 2008;**4**:23

[226] Job M, Agina S, Dapiya H. Occurrence of aflatoxigenic fungi in smoke-dried fish sold in Jos Metropolis. British Microbiology Research Journal. 2015;**11**(1):1-7. Article no. BMRJ. 21465

[227] Farag HEM, El-Taiby AA, Hassan HM. Assessment of ochratoxin A and aflatoxin B_1 levels in the smoked fish with special reference to the moisture and sodium chloride content. Research Journal of Microbiology. 2012;**6**(12):813-825

[228] Sun W, Han Z, Aerts J, Nie D, Jin M, Shi W, et al. A reliable liquid chromatography–tandem mass spectrometry method for simultaneous determination of multiple mycotoxins in fresh fish and dried seafoods. Journal of Chromatography. A. 2015;**1387**:42-48

Chapter 4

Mycotoxins: The Hidden Danger in Foods

Aycan Cinar and Elif Onbaşı

Abstract

Mycotoxins are secondary metabolites synthesized by a variety of fungal species such as *Aspergillus*, *Penicillium*, *Fusarium*, and *Alternaria*. These secondary metabolites are toxic and have a significant impact if they enter the production and food chain. Mycotoxins have attracted worldwide attention because of their impact on human health, huge economic losses, and domestic and foreign trade. Although more than 400 mycotoxins have been identified, most studies have focused on aflatoxins (AF), ochratoxin A (OTA), *Fusarium* toxins, fumonisin (FUM), zearalenone (ZEA), trichothecenes (TCT), and deoxynivalenol/nivalenol due to food safety and economic losses. This chapter will be addressing the type of mycotoxins, its importance in food industry, preventive measures, and implementation of hazard analysis critical control point (HACCP) to control mycotoxin.

Keywords: mycotoxin, aflatoxins, ochratoxin A, *Fusarium* toxins, fumonisin, zearalenone, trichothecenes, deoxynivalenol/nivalenol, food industry, HACCP

1. Introduction

Mycotoxins are secondary toxic metabolites with a wide variety of chemical structures synthesized by fungi (mold) [1]. Mycotoxins are thought to be a kind of "chemical defense system" to protect mold from insects, microorganisms, nematodes, grazing animals, and humans [2]. Molds reproduce by means of spores, and their small molecular weight spores are easily disseminated to environment by wind. They cannot be affected by the adverse environmental conditions and can be present in the latent state for long periods. Moreover, when the environmental conditions are appropriate, spores return to vegetative form and can form into new mold colonies. Agricultural products can be contaminated with mold in pre-harvest via insect and bird damage and harsh weather condition damage such as hail damage. In addition, selected harvesting method is one of the most important reasons in contamination of the mold to the products. Improper storage, transport, and marketing can also cause the mold growth and synthesis of mycotoxins [3].

Mycotoxin can occur in food and agricultural products via many contamination pathways, at any stage of production, processing, transport, and storage (**Figure 1**) [4]. Factors that affect mold growth and mycotoxin production are temperature, relative humidity, fungicides and/or fertilizers, interaction between the colonizing toxigenic fungal species, type of subtract and nutritional factors, geographical location, genetic requirements, and insect infestation [5, 6].

Approximately 400 fungal secondary metabolites are known to be toxic, and one quarter of agricultural products have been reported to be contaminated with

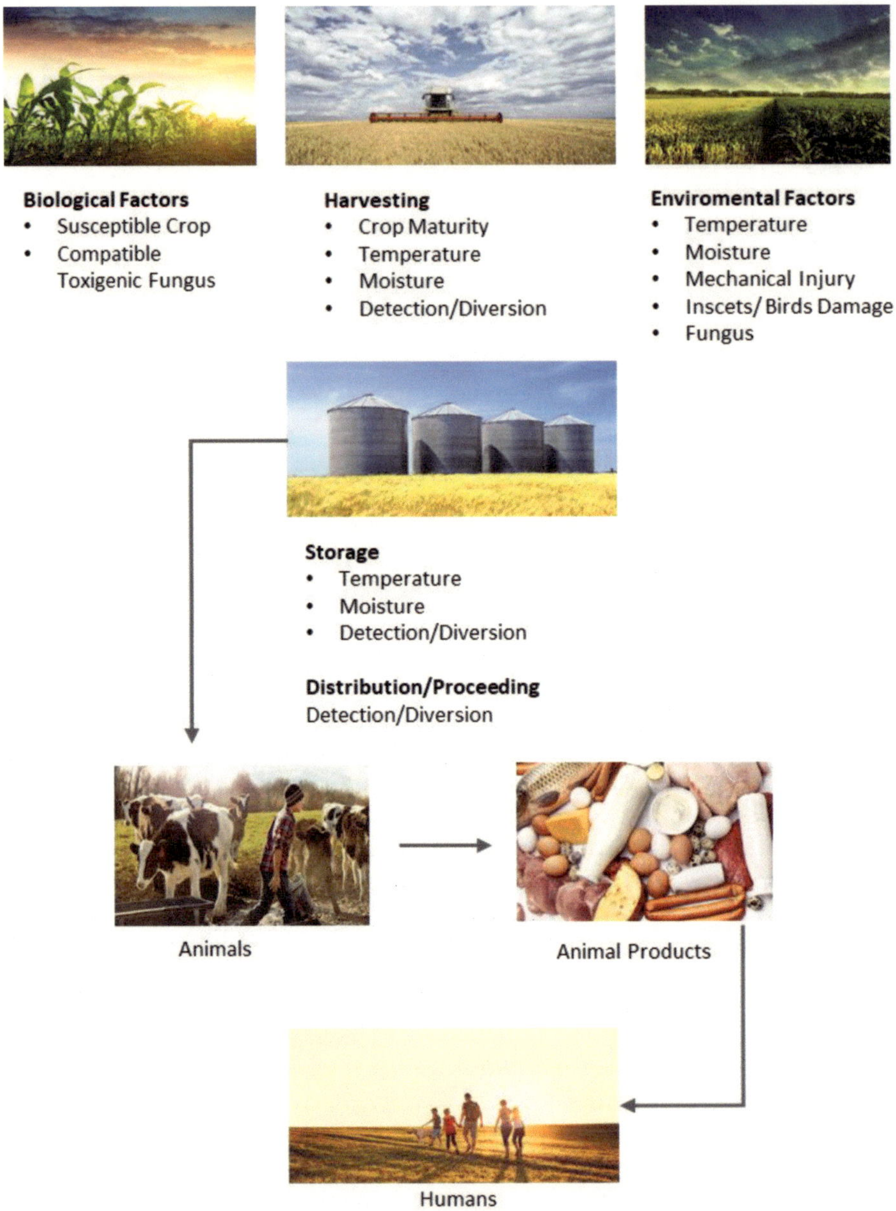

Figure 1.
Factors affecting mycotoxin occurrence in the food and feed chain [7, 8].

mycotoxins in the world [5–9]. While a type of mold may form more than one mycotoxin, a mycotoxin can be synthesized by many molds. The most common types of mold which are known to produce mycotoxins are *Aspergillus*, *Penicillium*, *Fusarium*, and *Alternaria* [10].

According to the result of many studies in poultry and mammals, mycotoxins can be carcinogenic, mutagenic, teratogenic, hepatotoxic, nephrotoxic, immunosuppressive, and embryotoxic [11]. The phenomenon of toxicity is called mycotoxicosis occurring after consumption of mycotoxin-contaminated product by human and animal [12].

Especially cereals, grains, nuts, oilseeds, fruits, dried fruits, vegetables, cocoa and coffee beans, wine, beer, herbs, and spices are major mycotoxin vectors since they are consumed by a large mass of people and animals [4]. Mycotoxins cause different degrees of toxicity according to exposure time, mycotoxin amount, physiological state, and sensitivity of the organism in humans and animals.

In addition to risk of public health, mycotoxins generate high level of economical loses for food industry due to reduced crop yields, lost trade revenues (local and international), and livestock illnesses [13, 14]. Elimination of mycotoxin is quite though due to resistant to physical, chemical, and biological methods; however, some of the measures described in the following sections may help to prevent mycotoxin. The methods used for mycotoxin determination are chromatography such as high-performance liquid chromatography (HPLC), thin-layer chromatography (TLC), gas chromatography-mass spectrometry (GC-MS), and also enzyme-linked immunosorbent assay (ELISA) technique and biosensor-based screening methods [15]. Detection is complicated due to limitations in analytical methodology [16]. Therefore, prevention of mold contamination and mycotoxin synthesis is essential for food safety in food industry.

According to the Food and Agricultural Organization (FAO), 77 countries have established guidance and regulations on mycotoxin in food and feed to control the level of mycotoxin. On the other hand, 13 countries including African countries still do not have specific regulation for food safety [4].

2. Importance of mycotoxin in food industry

Ergotism is one of the oldest determined mycotoxicoses (disease) in human and results from consumption of the ergot body in rye or other grains infected by a parasitic fungus of the genus *Claviceps*. The history of this disease is based on the outbreak of Spartans in 430 BC [17]. The world has been met with mycotoxin term after an extraordinary death of nearly 100,000 turkeys in near London, England, in 1960 due to a peanut (groundnut) meal imported from Brazil, contaminated with secondary metabolites from *Aspergillus flavus* (aflatoxins) [18]. Scientists focused on the occurrence and toxicology of mold metabolite that could cause serious health and economic losses after this case. Aflatoxin (AF) is the term derived from the name of one of the molds that produces it, *Aspergillus flavus*. Mycotoxins have been affecting people since 1960, which is the time of the finding of mycotoxin, and this problem still persists worldwide.

Mycotoxins can occur in the food in several ways (**Figure 1**), but technically divided into two groups; first is mold growth as a pathogen plant in field, another one is grow on stored. After plant materials are contaminated with mold spores from soil and air, they easily contaminate other food source, production area, laboratory, and even kitchen of our homes. Certain species of mold are capable of mycotoxin synthesis; therefore, each food contaminated with mold always may not contain mycotoxins. Nevertheless, moldy products are considered to be risky products in terms of mycotoxin.

Mycotoxins appear in almost all kinds of animal feed and products such as wheat bran, noug cake, pea hulls, maize grain, milk and meat, and also human food such as cereal, fruit and vegetables, spice, etc. [5]. Consuming these foods creates serious health risks in human and all animal species. Mycotoxin intake by feed or food causes chronic intoxication rather than acute symptoms. Acute toxicity is observed in high-dose mycotoxin exposure, and symptoms show a rapid effect such as borborygmy, abdominal pain, diarrhea, etc. On the other hand, low-level mycotoxin exposure in long period causes serious impairments in the liver, kidney, and immune system organs and tissues. Therefore, mycotoxin plays a significant role in cancer in these organs [2]. Some important mycotoxin health effects are shown in **Figure 2**. Toxic effects on humans and animals of important mycotoxins are shown in **Table 1** [19].

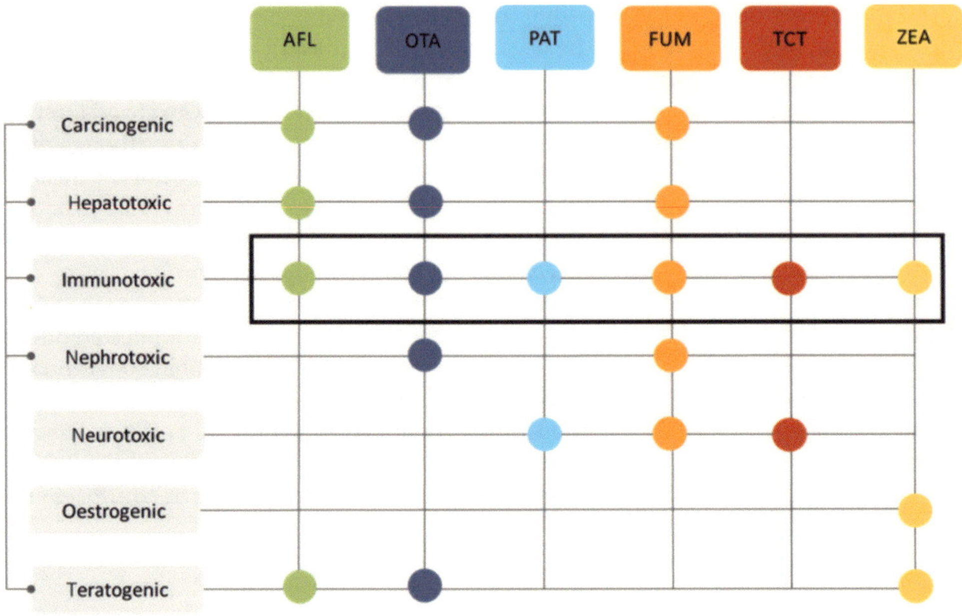

Figure 2.
Aflatoxin (AFL), ochratoxin A (OTA), patulin (PAT), fumonisin (FUM), trichothecenes (TCT), and zearalenone (ZEA) mycotoxin health effects [20].

Mycotoxins have caused many serious outbreaks worldwide. There was an outbreak that occurred in 1967, and 26 people were poisoned because of the consumption of moldy rice for up to 3 weeks in Taiwan [21]. An outbreak of aflatoxicosis affecting humans, reported in India, led to the death of 100 people in 1974 [22]. Another outbreak was reported in India in 1995, affecting 1424 people due to sorghum and maize contaminated with fumonisin [23]. During January–June 2004, an aflatoxicosis outbreak in eastern Kenya resulted in 317 cases and 125 deaths [24].

Mycotoxin contamination in foods and fodder has been becoming a global concern day by day. According to Food and Agricultural Organization (FAO) reports, it is estimated that mycotoxin affects nearly 25% of the world's crop each year and is causing huge agricultural product and industrial losses in billions of dollars [25]. For example, estimated annual loss in the United States is approximately $ 0.5–1.5 billion [19]. The main effects of mycotoxins on national economies can be thought in five ways:

1. Product yield losses due to toxigenic mold diseases

2. Decrease in commercial value because of contaminated food and feed

3. Human and animal health losses due to harmful impacts associated with mycotoxin-contaminated food and fodder consumption

4. Cost of analysis of mycotoxin

5. Strategies to control mycotoxin contamination

Economic impacts are felt by agricultural chain such as manufacturer of plant and animal, especially cereal industry, consumers, and briefly all farm-to-fork steps.

Mycotoxins	Genus/species	Major food	Toxic effects and diseases
Aflatoxin	*Aspergillus flavus* *A. parasiticus* *A. nomius* *Penicillium*	Cereals, feeds, oilseeds and pulp, coconut	Carcinogenic, hepatotoxicity, teratogenicity, decreasing immune systems, affecting the structure of DNA, hepatitis, bleeding, kidney lesions
Fumonisin	*Fusarium verticillioides* *F. culmorum*	Cereals, corn	Encephalomalacia, pulmonary edema, carcinogenic, neurotoxicity, liver damage, heart failure, esophageal cancer in humans
Ochratoxin OTA	*Aspergillus* *Penicillium* *A. ochraceus* *P. nordicum* *P. verrucosum*	Cereals, herbs, oil seeds, figs, beef jerky, fruits, and wine	Kidney and liver damage, loss of appetite, nausea, vomiting, suppression of immune system, carcinogenic
Patulin	*Aspergillus terreus* *A. clavatus* *Penicillium* *Penicillium carneum* *P. clavigerum* *P. griseofulvum*	Silage, wheat, feeds, apples, grapes, peaches, pears, apricots, olives, cereals	Neural syndromes, brain hemorrhage, skin lesions, skin cancer, lung, mutagenicity, antibacterial effect
Trichothecenes (T2, DON, DAS, HT2)	*Fusarium* *Cephalosporium* *Trichoderma* *Fusarium oxysporum*	Cereals, feeds, silage, legumes, fruits, and vegetables	Immune suppression, cytotoxic, skin necrosis, hemorrhage, anemia, granulocytopenia, oral epithelial lesions, GIS lesions, hematopoietic, alimentary toxic aleukia (ATA), hypotension, coagulopathy
Zearalenone	*Fusarium* *F. graminearum* *F. culmorum*	Cereals, corn, silage, timothy grass, fodder	Carcinogenic, hormonal imbalance estrogenic effect, reproductive problems, teratogenic

Table 1.
Name of some important mycotoxin-producing fungi, susceptible foods, and mycotoxin effects on humans and animals [19].

3. Worldwide important mycotoxin in food industry

3.1 Aflatoxins (AF)

Aflatoxins are a group of toxic secondary metabolites of filamentous fungi, *Aspergillus flavus*, *A. nomius*, and *A. parasiticus*, and the most important mycotoxins in the world for human food and animal feed [26]. On the other hand, recent studies have showed that *A. nomius*, *A. sergii*, *A. bombycis*, *A. minisclerotigenes*, *A. parvisclerotigenus*, *A. pseudocaelatus*, *A. pseudotamari*, and *A. ochraceoroseus* also have aflatoxigenic properties, but the occurrence of these species in nature is low [27]. The natural fungal multiplication subsequent to quantity of AF production is affected by various factors including environmental conditions (e.g., high temperature, moisture, and relative humidity), the presence of carbon dioxide and oxygen, mechanical damages, plant genre, insect infestation and amount of spores, and implementation of pesticides and fungicides [28, 29]. Among these, especially temperature and relative humidity are the most important effects of the formation and amount of AF as *A. flavus* has shown optimal growth at temperature from 29 to 35°C, maximum aflatoxin production at 24° C, and no production at temperatures below 13°C or above 42°C and relative humidity below 70% [30]. Heat processing, such as ultra-high-temperature (UHT) treatment, pasteurization, roasting, and baking, and also cold storage do not affect aflatoxin in foods since they are fairly stable and resistant [31, 32]. Approximately more than 14 various chemical forms of AF are present in nature; however, the most dangerous ones are aflatoxins B1, B2, G1, and G2 [33]. The nomenclature of aflatoxins with these letters is based on the color they exhibit under ultraviolet radiation (B, blue, and G, green) [34]. Various food products especially grown in hot and humid regions of the world are susceptible to fungal invasion and aflatoxin production, including groundnuts, maize, various spices, tree nuts, cottonseed, pistachios, copra, wheat, rice, etc. [25]. AFB1 is converted into metabolized AFM1 and excreted in milk in both human and lactating animals [35]. The European Commission, Codex Alimentarius Commission, Germany, Turkey, Switzerland, France, Sweden, Belgium, Argentina, Iran, and Honduras have regulated an acceptable limit for AFM1 at 50 ng/L for infants, for raw, pasteurized, and UHT milk. On the other hand, the United States, Brazil, China, Bulgaria, Czech Republic, Kuwait, and Serbia have accepted 500 ng/L level for AFM1 [31]. Aflatoxin contamination causes huge economic and critical health problem due to their high toxicity. For example, aflatoxin contamination is estimated to cause damages to the corn industry in the United States ranging from US $ 52.1 million to US $ 1.68 billion [36]. They are carcinogenic, hepatotoxic, and teratogenic, decrease immune systems, poison the body through respiratory, and can directly affect the structure of DNA [37]. Of all the human health effects associated with aflatoxin exposure, the weight of evidence is strongest for aflatoxin-related liver cancer and secondarily of the synergism between aflatoxin exposure and chronic HBV infection in liver cancer risk [38]. In 1974, there was an outbreak of hepatitis due to aflatoxin in India, resulting in an estimated 106 deaths [22]. In 2004 the largest outbreak was ever recorded, where 317 people became ill and 125 people died because of consumption moldy maize which early harvested and stored improper harvested condition [39]. In 2013, countries in Europe, including Romania, Serbia, and Croatia, reported that nationwide milk was contaminated with aflatoxin [40].

3.2 Ochratoxin A

Ochratoxin A (OTA) is a natural mycotoxin produced mainly by fungal type of *Aspergillus* and *Penicillium* under optimum environmental conditions and storage especially tropical and subtropical regions such as Eastern and South Europe, Canada, and South America [41, 42]. There are three types of ochratoxins, namely A, B, and C. Especially, OTA is known as the most common and important one for public and animal health. Although people are exposed to OTA by inhalation or dermal contact, various foods are the main source of exposure to OTA including maize, sorghum, wheat, rice, barley, rye, bread, oats, flour, pasta, grapes, infant cereals, apples, peaches, strawberries, pears, oranges, figs, mangoes, wine, tomatoes, coffee beans, watermelons, nuts, rapeseed, sesame seeds, spice, soybeans, cocoa, peanuts, chickpeas, milk and milk-based baby formulae, eggs, cheese, yam, potatoes, garlic, onions, fish, pork, poultry, jerky, and dried beans [43]. Recently, the presence of OTA has been detected in bottled water [44], plant food supplement, and food coloring agent [45]. According to the European Commission report, the estimated adult exposure to OTA is as follows: 44% cereals, 10% wine, 9% coffee, 7% beer, 5% cacao, 4% dried fruits, 3% meat, 3% spices, and 15% others [46]. For the first time in 1970, the presence of OTA was detected in human blood in Balkans [47]. In the review of Malir et al., published data on OTA in human blood samples from healthy persons were compiled, and concentrations higher than 1.0 g/L were observed in several countries [48].

Huge amount of economic losses occurs resulting from OTA contamination on feed and food particularly livestock production. Exposure of OTA causes renal dysfunction (suspected in Balkan endemic nephropathy) and also is considered to be teratogenic, immunotoxigenic, nephrotoxic, carcinogenic, embryotoxic, hepatotoxic, and especially nephrotoxic in laboratory and farm animals [43, 49].

3.3 *Fusarium* toxins

Fusarium toxins are secondary metabolites synthesized by toxigenic molds including *Fusarium oxysporum*, *F. culmorum*, *F. roseum*, and *F. graminearum* [50]. Fumonisins (FBs), zearalenone (ZEA), trichothecenes, deoxynivalenol (DON), and nivalenol (NIV) are the most common *Fusarium* mycotoxin groups [51]. Recently fusaproliferin (FUS), beauvericin (BEA), enniatins (ENNs), and moniliformin (MON) are discovered but less studied [52]. *Fusarium* disease outbreak on cereal products such as wheat, barley, and maize causes worldwide economic losses due to yield loss and reduced grain quality, for example, losses in the United States of $ 1–20 million in a normal year and $31–46 million in a year [53]. *Fusarium* mycotoxin has both acute and chronic toxic effects and been shown to cause a wide variety of toxic effects in animals [54]. Spontaneous outbreaks of *Fusarium* mycotoxicosis have been reported in Europe, Asia, Africa, New Zealand, and South America. Moreover, chronic intake of these mycotoxins is reported on a regular and more widespread basis due to their global occurrence [55]. *Fusarium* mycotoxin limits specified in unprocessed cereals, milling products, and cereal foodstuffs are 200–1750 μg/kg for DON, 20–400 μg/kg for ZEN, and 200–4000 μg/kg for the sum of B1 + B2 fumonisins (FB1 + FB2 combined) according to the European Commission (19 December 2006).

3.3.1 Fumonisin

Fumonisins are generated by various fungal species such as *Fusarium verticillioides* and *F. proliferatum* also by *A. niger* and were discovered in 1988 in South Africa [56, 57]. Nowadays 28 types of fumonisin have been identified that are divided into four groups, fumonisins A (A1, A2, and A3), fumonisins B (B1, B2, and B3), fumonisins C (C4, C3, and C1), and fumonisins P (P1, P2, and P3), but the most important group of fumonisins is the B group, which contains fumonisins B1 (FB1), B2 (FB2), B3 (FB3) [58].

The International Agency for Research on Cancer (IARC) identified FB1 as possibly carcinogenic to humans (group 2B). Recent studies reported that FB1 causes an increased prevalence of esophageal and liver cancer in humans [59]. Furthermore, this mycotoxin has been found to have toxic effects against several organs (nervous and cardiovascular systems, liver, lung, kidney) in animals [60]. Fumonisins are largely found in corn and corn-based foods and also FB1 in rice, beer, sorghum, cowpea seeds, triticale, beans, asparagus, and soybeans [61].

3.3.2 Zearalenone (ZEA)

Zearalenone (ZEA), known as an estrogenic mycotoxin, is a secondary metabolite produced by *Fusarium* species such as *F. graminearum*, *F. culmorum*, *F. cerealis*, *F. equiseti*, *F. crookwellense*, and *F. semitectum* (mainly *F. culmorum* and *F. graminearum*) [62]. The main contamination source of ZEA is cereal-based foods such as maize, sorghum, wheat, rice, barley, oats, and also nuts, soybean, and sesame [63].

Several in vivo studies found that ZEA disrupts hormonal balance due to its similarity to naturally occurring estrogens [64]. The mycotoxin has high affinity for estrogen receptors, causing reproduction and fertility disorders in mammals [65]. In addition, it is known that progressive exposure to endocrine-modulatory compound has been linked with carcinogenesis in human [64]. According to the European Food Safety Authority (EFSA) report in 2014, the bioavailability of toxin is up to 80% in human and animals such as rats, rabbits, and pigs [66]. Moreover, recent works report ZEA is metabolized in the liver and has shown hepatotoxic, immunotoxic, carcinogenic, and nephrotoxic effect in animal tests [67–69]. As this mycotoxin possesses such consumer health risks, the European Union (EU) has prescribed the limits of ZEA (20–350 μg/kg) for various processed and unprocessed cereals [66].

3.3.3 Trichothecenes (TCT)

Trichothecenes are a large group of mycotoxins produced predominantly by *Fusarium* species although produced by other fungal genera such as *Trichoderma*, *Trichothecium*, *Stachybotrys*, *Verticimonosporium*, *Cephalosporium*, *Myrothecium*, and *Cylindrocarpon* spp. [70]. More than 200 different trichothecenes and trichothecene derivatives have been isolated. Trichothecenes are classified into four types (A–D). Type A and type B are the most prevalent type occurring widely in cereals [71, 72]. Type A trichothecenes such as T-2 and HT-2 toxins, diacetoxyscirpenol (DAS), monoacetoxyscirpenol (MAS), and neosolaniol (NEO) are synthesized mainly by *F. sporotrichioides* and *F. langsethiae*. On the other hand, type B including deoxynivalenol (DON), the co-contaminants 3- and 15-acetyl DON (3A-DON or 15A-DON), and fusarenon-X (FUS-X; synonym 4-acetylnivalenol) are mainly produced by *F. graminearum* and *F. culmorum* [73]. Moreover, another important type B member, nivalenol (NIV), is commonly synthesized by *F. poae* in cereals [74].

The mechanism of action of trichothecenes is based on the inhibition of protein synthesis in eukaryotes. This mycotoxin affects peptidyl transferase enzyme binding the 60S ribosomal subunit, thus causing the inhibition of protein translation and ribotoxic stress [75]. Also, Pestka reported these groups of mycotoxins cause immunosuppression or immune stimulation by affecting the leucocytes [76].

The family of trichothecenes has a significant impact on cereal and grain production due to health risk for human consumption, livestock feed, or malting purposes [77, 78]. According to report from the FDA, economic losses associated with mycotoxin ranges from USD 0.5 million to over USD 1.5 billion from aflatoxin (corn and peanuts), fumonisin (corn), and deoxynivalenol (wheat) in the United States. [72]. Hence, control of these mycotoxins is essential for human and animal health and economic reasons.

3.3.3.1 Deoxynivalenol/nivalenol

Deoxynivalenol (DON), known as vomitoxin, is the most commonly detected trichothecenes in grains such as wheat, barley, oats, rye, and corn and less often in rice, sorghum, and triticale [79]. Even though NIV presence of cereals appears generally to be lower than DON [80], it has been reported that the occurrence of NIV in of wheat and barley is as prevalent as that of deoxynivalenol (DON) in Japan [81]. According to animal toxicity studies, NIV shows higher toxicity than DON. The LD50 values for DON and NIV in tests in mice were 78 and 39 mg/kg, respectively, and DON and NIV, similarly to other trichothecenes, show inhibitor effect on cell metabolism such as protein, DNA, and RNA synthesis [82]. In addition, these mycotoxins affect cell division and mitochondrial functions [83, 84, 70]. Both mycotoxins exhibit major symptoms such as abdominal discomfort, diarrhea, vomiting, and inflammation of the throat, weight loss, and anorexia [85].

The World Health Organization (WHO) reported that trichothecenes shows fatal and chronic intoxications on human and livestock and also DON shows teratogenic, neurotoxigenic, and immunosuppressant effects [86].

According to the conducted BIOMIN World Mycotoxin Survey, DON appeared in 81% of livestock feed from 81 countries worldwide followed by fumonisins that were detected in 71% of samples. Therefore, DON is reported as the most common mycotoxin worldwide (https://www.biomin.net/en/biomin-mycotoxin-survey/).

4. Management of mycotoxin prevention

Food safety is a key component in public health issue, and a mycotoxin is a huge food safety risk in developing countries. Prevention is the most important and effective way in reducing fungal growth and mycotoxin production to ensure food safety. The following steps that explain prevention and control of mycotoxin occurrence include good agricultural practices (GAP) in field, control practices of harvesting and storage, physical methods (cleaning, milling, etc.), implementation of biotechnological application, biological control through the use of controlled atmosphere during storage, detoxification/degradation, and fermentation techniques.

Pre-harvesting is considered first and one of the most important stages to prevent mold growth and mycotoxin synthesis. Several strategies are available for the produce of healthy products and reduce the mold formation at pre-harvesting, including selection of plants according to the soil structure and production capacity, use of plant which is resistant to fungi and insects, irrigation time, make fertilization, use of insecticides to prevent insect damage [87].

Harvesting at the appropriate time periods (low moisture and full maturity) is essential for reducing the risk of a mycotoxin contamination since overmaturity creates sensitivity to mold growth. Additionally, suitable harvesting equipment and procedures should be used, and crops should be dried after maturity to both reduce grain moisture to safe levels [88].

The latest technological advances provided new paths in mycotoxin control strategies that include the use of a controlled atmosphere with inhibitory or a protective effect and use of naturally occurring compounds under different conditions and essential oils with antioxidant properties to decrease fungal growth and mycotoxin production in grains during storage [89]. Moreover, these strategies also include using regularly cleaned transport vehicles to prevent cross contamination of products; monitoring of temperature, humidity, aeration and pest infestation periodic during storage [90]; using mold inhibitors (propionic acid) to contaminated food and feed; and application of disinfectant such as sodium hypochlorite to storage area [91].

Some studies have shown that using physical methods (dehulling, washing, sorting, and cleaning of visible moldy seed) reduces different mycotoxin species in foods regardless of grain genre [70]. Scudamore and Pascale et al. [92] and Patel [93] observed a reduction of T-2 (62%) and HT-2 (53%) and DON (50%) in wheat seeds after cleaning. Scudamore and Patel also reported a 32% reduction in fumonisin levels in corn in an industrial enterprise [94]. Moreover, milling is an important effect in the reduction of *Fusarium* mycotoxins in grains especially wet milling of maize which has shown to result in the degradation of mycotoxins [95].

One of the best applicable strategies for the prevention of mycotoxin formation is the cultivation of fungal infestation-resistant plants and improvement of the genetic composition to suppress mycotoxin production [96]. The benefits of biotechnological applications were observed with Aflasafe. Aflasafe is a biocontrol product that includes a blend of four fungal species covered over grains which reduce aflatoxigenic fungi that produce AFs in maize and groundnuts (https://aflasafe.com/).

Mycotoxins are resistant to heat and cannot be completely destroyed under normal cooking process. On the other hand, mycotoxin reduction has been determined after heating, and this may be the result of reactions changing the chemical structure [70]. Ryu et al. reported heat treatment (at temperature 120–160°C) causes a reduction between 66 and 83% of ZEN [97]. Scott and Lawrence also reported a reduction of 60–100% of fumonisins with a heat treatment at 190°C (60 min) and 220°C (25 min).

Biological control of mycotoxins via detoxification/degradation offers a promising alternative method [98]. Recently the effectiveness of fermentation for the reduction and elimination of mycotoxins has also been proven. Studies documented in the literature generally show that mycotoxins are reduced by conversion, detoxification, binding, degradation, and decontamination after food fermentation [99]. Modification of the chemical structure of the mycotoxin molecule, removal or detoxification/inactivation, and adhesion to bacterial cell walls provide a reduced toxicity during fermentation [99]. Implementation of these preventive methods cannot solve the problem alone; also it must be an integral part of an integrated food safety management system based on the hazard analysis and critical control point (HACCP).

5. Implementation of HACCP to mycotoxin control

HACCP is a food management system where food safety is addressed through the analysis, control, and monitoring of physical, chemical, and biological hazards from raw material manufacturing, supply, and handling to production, distribution, and consumption of the finished product [100]. The National Advisory Committee on Microbiological Criteria for Foods (NACMCF) published a guideline about HACCP containing seven basic principles, decision tree, and all plans in 1992 [101]. Implementation of HACCP is an effective strategy for prevention, control, and periodic monitoring of mycotoxin in all stages from field to the consumer. There are 12 successive steps recommended to implementation of HACCP system. Previous HACCP studies can be researched to set up tasks from 1 to 5 that specify each food process, and tasks required for mycotoxin control begin at 6 (Principle 1).

1. Establish the HACCP team.

2. Describe the product.

3. Identify the product's intended use.

4. Draw up the commodity flow diagram.

5. Confirm the flow diagram on-site.

6. Identify and analyze hazard(s) (Principle 1).

7. Determine the critical control points (CCPs) (Principle 2).

8. Establish critical limits for each (CCP) (Principle 3).

9. Establish a monitoring procedure (Principle 4).

10. Establish corrective action (Principle 5).

11. Verify the HACCP plan (Principle 6).

12. Keep record (Principle 7).

Principle 1: identify and analyze hazard—food safety hazards for HACCP programs are divided into three groups: biological (bacteria, viruses, parasites, etc.), chemical (cleaning agents, pest control, pesticides, biocides, mycotoxin), and physical (glass or metal fragments, jewelry, etc.). Mycotoxins are identified as biological hazards because they are secondary metabolites of mold and also identified as a chemical hazard that appears as residues in food.

Principle 2: determine critical control points (CCPs)—determining CCPs is an essential step which is decided using the HACCP decision tree to eliminate or prevent a food safety hazard or reduce it to an acceptable level. Dried figs and other dried fruits, pistachios and other edible nuts and cereals, and also animal feed such as maize, groundnut cake, cottonseed cake, babassu, palm kernel cake, copra cake, etc. are susceptible to mycotoxin in planting, harvesting, production, storage, and transport according to EC regulations. Mycotoxins can be considered a CCP for these products. For example, *Aspergillus flavus* is a CCP in maize production. It is a pathogenic fungus which colonizes in broken kernels in stored maize. High concentration of aflatoxin can cause public health problem, rejection of the final product or product recalls, litigation, etc. [102].

Principle 3: establish critical limits—critical limits must be defined and verified for each CCP. Mycotoxin acceptable limits can be set by country regulation and customer or producer specification which is below of the regulatory mycotoxin limit (**Table 2**).

Principle 4: establish a monitoring system for each CCP—identifying an appropriate, sensitive, and rapid monitoring method which applies physical, chemical, and biological measurement or observations for each critical control point. HPLC, GC, ELISA, OWLS-based biosensors, rapid test kits, etc. are used to detect mycotoxin level.

Principle 5: establish a corrective action—Corrective action must be established when monitoring result indicates that there is a deviation of target CCP

Crops and tolerated levels of mycotoxins (μgkg^{-1})					
Country	**Mycotoxins**	**Rice**	**Maize**	**Spices**	**Fruit juices**
Brazil	AFB1/AFG1	30	30	30	30
China	AFB1	10	20	—	—
France	FB1	1000	1000	—	—
Hungary	Total AF OTA	50 5	50 5	- -	- -
Japan	AFB1 Patulin	10	10	10	—
The United States	Total AF Patulin	20 -	20 -	20 -	- 50
Turkey	AFB1 Patulin	2 -	2 -	5 -	- -

Table 2.
Global regulation of mycotoxin contamination in agricultural products [103].

Step/CCP	Hazard analysis		Monitoring			Corrective action
	Hazard	Control	Critical limit	Monitoring	Frequency	
Pre-harvest/ growing	Low soil moisture leading to plant stress during kernel development	Irrigate	Lower limit of critical water activity (aw) (check with your agronomist/extension staff for an exact value)	Measure soil moisture and record	Weekly on Monday morning	Additional irrigation; record amounts
	Insufficient soil nutrients leading to plant stress during kernel development	Fertilize	N, P, and K applications as recommended for hybrid by local agronomists (insert the values)	Fertilizer applied (appropriate for soil type and hybrid); amounts and type recorded	As recommended for hybrid	Additional fertilizer; record amounts added
	Insect attack leading to damaged kernels	Integrated pest management (IPM) plan	Insect population within acceptable limits as determined by control program	Visual inspection and sample, with results recorded	Weekly	Apply pesticide in accordance with IPM plan
Harvest	Damage to kernels from harvester	Harvest when kernels are dry	Moisture content ≤14%	Measure and record grain moisture	Prior to harvest	Delay harvest till kernels are dried enough
Storage	Excessive moisture content of kernels	Do not store until kernels are dry	Moisture content ≤14%	Measure and record grain moisture	Immediately prior to storage	Dry mechanically
	Insect attack, allowing fungi to penetrate kernels	IPM plan	No evidence of insect or rodent infestation using inspection protocols specified in IPM plan	Visual inspection with results recorded	Weekly	Apply pest control methods in accordance with IPM plan
	High ambient humidity and temperature	Aerate grain to control temperature and humidity	Temperature and humidity within limits recommended in industry literature	Measure and record humidity, ambient temperature, and airflow	Daily during storage	Adjust aeration time of day or airflow to achieve desired temperature and humidity

Table 3.
HACCP plan of maize [102].

value. Taking appropriate corrective actions immediately is essential to producing safe food [103]. Corrective actions must ensure that the CCP is taken under control. Corrective action sample of maize production is given in **Table 3**.

Principle 6: establish verification procedures—regularly at the specified intervals, it must be verified by checking whether the levels of mycotoxin in the final product are within acceptable levels. The following steps are used for verification:

- Microbiological and/or chemical tests can be used to confirm which product is meeting CCP.
- Asking questions especially to CCP employees.
- Internal or external audit by independent person to check whether HACCP system is being implemented.

Principle 7: establish documentation and record keeping—record keeping is an evidence of how you identify, monitor, and verify each hazard. HACCP plan, flowchart of product, product description, HACCP team, hazard analysis documents, analysis result sheet, etc. are required for monitoring whether control of each hazard is appropriate or not.

6. Conclusion

Mycotoxin is a well-known food safety risk, which is a threat to human and livestock health, and has high economic significance in food industry. Recently, the food industry has become aware of the new term modified mycotoxins introduced by Rychlik et al. (masked mycotoxin) [104]. Food safety risk has risen since masked mycotoxins which pose many difficulties including the unknown occurrence/co-occurrence of these compounds and their toxicological properties. In addition, Lorenz et al. reported that the European Food Safety Authority (EFSA) has taken into account efforts to address this emerging issue in food safety by developing strategies on how to evaluate potential added health risk due to the occurrence of modified mycotoxins [104].

Mycotoxigenic molds are difficult to prevent and control due to their widespread presence in nature. Prevention of mycotoxin synthesis in all stages of food processing is an essential point for public health and economic reasons. Many practices used for prevention of mycotoxin include good agricultural practices (GAP) in field, control practices of harvesting and storage, physical methods (cleaning, milling, etc.), implementation of biotechnological application, biological control through the use of controlled atmosphere during storage, detoxification/degradation, and fermentation techniques.

Meanwhile a number of techniques for mycotoxin control and management prove to be quite costly and/or unenforceable in some cases. On the other hand, using fermentation process for appropriate process has been recommended for mycotoxin reduction by Adebiyi et al. [99]. In the future, more emphasis should be given to nanotechnology and genetic engineering practices in the development of durable product types to ensure food safety.

In addition to these applications, food safety management systems such as HACCP, GAP, and good manufacturing practices (GMP) should be integrated at all stages of production, transport, and storage, in order to minimize contamination in food industry. Also fairly new food safety system including threat assessment critical control points (TACCP), vulnerability critical control points (VACCP), and hazard analysis and risk-based preventive controls (HARPC) should be investigated and implemented to ensure an effective control system.

Author details

Aycan Cinar* and Elif Onbaşı
Faculty of Engineering and Natural Sciences, Department of Food Engineering, Bursa Technical University, Bursa, Turkey

*Address all correspondence to: aycan.cinar@btu.edu.tr

References

[1] Turnera NW, Subrahmanyam S, Piletskyb SA, et al. Analytical methods for determination of mycotoxin: A review. Anaytica Chimica Acta. 2009;**632**:168-180. DOI: 10.1016/j.aca.2008.11.010

[2] Etzel RA. Indoor mold and children's health. Environmental Health Perspectives. 1999;**107**(Suppl 3):463. DOI: 10.1289/ehp.107-1566224

[3] Forsythe SJ. The Microbiology of Safe Food. 2nd ed. Wiley-Blackwell: UK; 2010. 496 p

[4] Ferrão J, Bell V, Chabite IT, Fernandes TH. Mycotoxins, food and health mycotoxins, food and health. Journal of Nutritional Health & Food Science. 2017;**5**(7):1-10. DOI: 10.15226/jnhfs.2017.001118

[5] Tola M, Kebede B. Occurrence, importance and control of mycotoxins: A review. Food Science and Technology. 2016;**2**:1191103. DOI: 10.1080/23311932.2016.1191103

[6] Deligöz E, Bilge N. Sütle Gelen Tehdit: Aflatoksin M1. Türk Tarım - Gıda Bilim ve Teknoloji Dergisi. 2017;**5**(8):846-857. DOI: 10.24925/turjaf.v5i8.846-857.1111

[7] Değirmencioğlu N, Eseceli H, Çokal Y, Bilgiç M. From safety feed to safety food: The application of HACCP in mycotoxin control. Archiva Zootechnica. 2005;**8**

[8] CAST. Mycotoxins: Risks in Plant, Animal, and Human Systems. Council for Agricultural Science and Technology: Ames IA; 2003

[9] Escrivá L, Font G, Manyes L, Berrada H. Studies on the presence of mycotoxins in biological samples: An overview. Toxins. 2017;**9**(8):251. DOI: 10.3390/toxins9080251

[10] World Health Organization. Mycotoxins; 2018. Available from: https://www.who.int/news-room/fact-sheets/detail/mycotoxins

[11] Maia F, Freire F, Guedes MIF. Mycotoxins and their effects on human and animal health. Food Control. 2014;**36**(1):159-165. DOI: 10.1016/j.foodcont.2013.08.021

[12] Peraica M, Radic B, Lucic A, Pavlovic M, et al. Toxic effects of mycotoxins in humans. Bulletin of the World Health Organization. 1999;**77**(9):754-766. Available from: https://www.who.int/bulletin/archives/77(9)754.pdf

[13] Abrunhosa L, Morales H, Soares C, Calado T, Vila-Chã AS, Pereira M, et al. A review of mycotoxins in food and feed products in Portugal and estimation of probable daily intakes. Critical Reviews in Food Science and Nutrition. 2016;**56**(2):249-265. DOI: 10.1080/10408398.2012.720619

[14] Zhu Y, Hassan YI, Watts C, Guelph TZ. Innovative technologies for the mitigation of mycotoxins in animal feed and ingredients—A review of recent patents. Animal Feed Science and Technology. 2016;**216**:19-29. DOI: 10.1016/j.anifeedsci.2016.03.030

[15] Bueno D, Istamboulie G, Munoz R, Marty JL. Determination of mycotoxins in food: A review of bioanalytical to analytical methods applied spectroscopy reviews. Applied Spectroscopy. 2015;**50**:728-774. DOI: 10.1080/05704928.2015.1072092

[16] Murphy PA, Hendrich S, Landgren C. Food mycotoxins: An update. Journal of Food Science. 2006;**71**:5. DOI: 10.1111/j.1750-3841.2006. 00052.x

[17] Bove FJ. The Story of Ergot. Basel, Switzerland: Karger; 1970. 297 p. DOI: 10.1159/isbn

[18] Zain ME. Impact of mycotoxins on humans and animals. Journal of Saudi Chemical. 2011;**15**:129-144. DOI: 10.1016/j.jscs.2010.06.006

[19] Oguz H. Mikotoksinler ve Önemi. Turkiye Klinikleri Veterinary Sciences-Pharmacology and Toxicology. 2017;**3**(2):113-119

[20] Egmond HV. Aflatoxins. 2013. http://www.seerural.org/wp-content/uploads/2013/04/2013-04-04_Aflatoxins-An-introduction-Risk-Assessment_Hans-van-Egmond_Wageningen.pdf

[21] Ling K, Wang JJ, Wu R, Tung TG, Lin SS, Lin TM. Intoxication possibly caused by aflatoxin B1 in the moldy rice in Shvangshih township. Journal of the Formosan Medical Association. 1967;**66**:729

[22] Krishnamachari K, Bhat RV, Nagarajan V, Tilac T. Investigations into an outbreak of hepatitis in Western India. The Indian Journal of Medical Research. 1975;**63**:1036-1048

[23] Raghavender C. Human and animal disease outbreaks in India due to mycotoxins other than aflatoxins. World Mycotoxin Journal. 2008;**2**(1):23-30. DOI: 10.3920/WMJ2008.1066

[24] Probst C et al. Outbreak of an acute aflatoxicosis in Kenya in 2004: Identification of the causal agent. Applied and Environmental Microbiology. 2007;**73**(8):2762-2764. DOI: 10.1128/AEM.02370-06

[25] Alshannaq A, Yu JH. Occurrence, toxicity, and analysis of major mycotoxins in food. International Journal of Environmental Research and Public Health. 2017;**14**(6):632. DOI: 10.3390/ijerph14060632

[26] Kurtzman CP, Horn BW, Hesseltine CW. *Aspergillus nomius*, a new aflatoxin-producing species related to *Aspergillus flavus* and *Aspergillus tamarii*. Antonie Van Leeuwenhoek. 1987;**53**(3):147-158

[27] Campagnollo FB, Ganev KC, Khaneghah AM, Portela JB, Cruz AG, Granato D, et al. The occurrence and effect of unit operations for dairy products processing on the fate of aflatoxin M1: A review. Food Control. 2016;**68**:310-329

[28] Jay JM. Microbiologia de Alimentos. Artmed: Porto Alegre; 2005

[29] Bryden WL. Mycotoxin contamination of the feed supply chain: Implications for animal productivity and feed security. Animal Feed Science and Technology. 2012;**173**:134-158

[30] Campagnollo FB, Ganev KC, Khaneghah AM, Portela JB, Cruz AG, Granato D, et al. The occurrence and effect of unit operations for dairy products processing on the fate of aflatoxin M1: A review. Food Control. 2016;**68**:310-329. DOI: 10.1016/j.foodcont.2016.04.007

[31] Bahrami R, Shahbazi Y, Nikousefat Z. Aflatoxin M1 in milk and traditional dairy products from west part of Iran: Occurrence and seasonal variation with an emphasis on risk assessment of human exposure. Food Control. 2016;**62**:250-256

[32] Picinin LCA, Cerqueira MMOP, Vargas EA, Lana ÂMQ, Toaldo IM, Bordignon-Luiz MT. Influence of climate conditions on aflatoxin M1 contamination in raw milk from Minas Gerais state, Brazil. Food Control. 2013;**31**:419-424

[33] Shahbazi Y. Aflatoxin M1 contamination in milk and dairy products: Implications on human health. Nutrients in Dairy and Their Implications for Health and Disease. 2017;**1**:237-250. DOI: 10.1016/B978-0-12-809762-5.00019-X

[34] Anfossi L, Baggiani C, Giovannoli C, D'Arco G, Passin C, Giraudi G. Occurrence of aflatoxin M1 in Italian cheese: Results of a survey conducted in 2010 and correlation with manufacturing, production season, milking animals, and maturation of cheese. Food Control. 2012;**25**:125-130

[35] Fallah AA, Rahnama M, Jafari T, Saei-Dehkordi SS. Seasonal variation of aflatoxin M1 contamination in industrial and traditional Iranian dairy products. Food Control. 2011;**22**(10):1653e1656

[36] Mitchell NJ, Bowers E, Hurburgh C, Wu F. Potential economic losses to the US corn industry from aflatoxin contamination. Food Additives & Contaminants. Part A, Chemistry, Analysis, Control, Exposure & Risk Assessment. 2016;**33**(3):540-550. DOI: 10.1080/19440049.2016.1138545

[37] Abyaneh RM, Chang PK, Ghahfarokhi MS, Rai M. Global health issues of aflatoxins in food and agriculture: Challenges and opportunities. Frontiers in Microbiology. 2014;**5**:420. DOI: 10.3389/fmicb.2014.00420

[38] Wu F. Global impacts of aflatoxin in maize: Trade and human health. World Mycotoxin Journal. 2015;**8**(2):137-142. DOI: 10.3920/WMJ2014.1737

[39] Azziz-Baumgartner E, Lindblade K, Gieseker K, Schurz Rogers H, Kieszak S, Njapau H, et al., The Aflatoxin Investigative Group. Case-control study of an acute aflatoxicosis outbreak – Kenya, 2004. Environmental Health Perspectives. 2005;**113**:1779-1783

[40] Abbas HK, Reddy KRN, Salleh B, Saad B, Abel C, Shier WT. An overview of mycotoxin contamination in foods and its implications for human health. Toxin Reviews. 2010;**29**:1-26. DOI: 10.3109/15569541003598553

[41] Yeni F, Yavaş S, Alpas H, Soyer Y. Most common foodborne pathogens and mycotoxins on fresh produce: A review of recent outbreaks. Critical Reviews in Food Science and Nutrition. 2016;**56**(9):1532-1544. DOI: 10.1080/10408398.2013.777021

[42] Chrevatidis A. Mycotoxins/occurrence and determination. In: Caballero B, editor. Encyclopedia of Food Sciences and Nutrition. 2nd ed. Maryland, United States: Academic Press; 2003. pp. 4089-4096. DOI: 10.1080/10408398.2013.777021

[43] Leitão AL. Occurrence of ochratoxin A in coffee: Threads and solutions. A mini-review. Beverages Journal. 2019;**5**:36. DOI: 10.3390/beverages5020036

[44] Mata AT, Ferreira JP, Oliveira BR, Batoreu MC, Barreto CMT, Pereira VJ, et al. Bottled water: Analysis of mycotoxins by LC-MS/MS. Food Chemistry. 2015;**176**:455-464

[45] Solfrizzo M, Piemontese L, Gambacorta L, Zivoli R, Longobardi F. Food coloring agents and plant food supplements derived from *Vitis vinifera*: A new source of human exposure to ochratoxin A. Journal of Agricultural and Food Chemistry. 2015;**63**:3609-3614

[46] EC. Assessment of Dietary Intake of Ochratoxin A by the Population of EU Member States; Report of the Scientific Cooperation, Task 3.2.7; Directorate-General Health and Consumer Protection. Rome, Italy: European Commission; 2002

[47] Hult K, Plestina R, Habazin-Novak V, Radic B, Ceovic S. Ochratoxin A in human blood and Balkan endemic nephropathy. Archives of Toxicology. 1982;**51**:313-321

[48] Malir F, Ostry V, Pfohl-Leszkowicz A, Malir J, Toman J. Ochratoxin A: 50 years of research. Toxins (Basel). 2016;**8**:191

[49] Klimke-B TR, Wu F. Ochratoxin, A and human health risk: A review of the evidence. Critical Reviews in Food Science and Nutrition. 2015;**55**(13):1860-1869. DOI: 10.1080/10408398.2012.724480

[50] Caldwell RW, Tuite J, Stob M, Baldwin R. Zearalenone production by *Fusarium* species. Applied Microbiology. 1970 Jul;**20**(1):31-34

[51] Ferrigo D, Raiola A, Causin R. *Fusarium* toxins in cereals: Occurrence, legislation, factors promoting the appearance and their management. Molecules. 2016;**21**:627. DOI: 10.3390/molecules21050627

[52] Summerell BA, Leslie JF. Fifty years of *Fusarium*: How could nine species have ever been enough? Fungal Diversity. 2011;**50**:135-144

[53] Wu F. Measuring the economic impacts of *Fusarium* toxins in animal feeds. Animal Feed Science and Technology. 2007;**137**:363-374. DOI: 10.3390/toxins6020430

[54] Escrivá L, Font G, Manyes L. In vivo toxicity studies of *Fusarium* mycotoxins in the last decade: A review. Food and Chemical Toxicology. 2015;**78**:185-206. DOI: 10.1016/j.fct.2015.02.005

[55] Cortinovis C, Pizzo F, Spicer LJ, Caloni F. Fusarium mycotoxins: Effects on reproductive function in domestic animals-a review. Theriogenology. 2013;**80**:557-564

[56] EFSA. Opinion of the scientific panel on contaminants in food chain on a request from the commission related to fumonisins as undesirable substances in animal feed. EFSA Journal. 2005;**235**:1-32. DOI: 10.2903/j.efsa.2005.235

[57] Frisvad JC, Larsen TO, Thrane U, et al. Fumonisin and ochratoxin production in industrial *Aspergillus niger* strains. PLoS One. 2011;**6**(8):e23496. DOI: 10.1371/journal.pone.0023496

[58] Escrivá L, Font G, Manyes L. In vivo toxicity studies of fusarium mycotoxins in the last decade: A review. Food and Chemical Toxicology. 2015;**78**:185-206. DOI: 10.1016/j.fct.2015.02.005

[59] Ueno Y, Iijima K, Wang SD, Sugiura Y, Sekijima M, Tanaka T, et al. Fumonisins as a possible contributory risk factor for primary liver cancer: A3-year study of corn harvested in Haiman, China by HPLC and ELISA. Food and Chemical Toxicology. 1997;**35**:1143-1150

[60] Bertero A, Moretti A, Spicer LJ, Caloni F. Fusarium molds and mycotoxins: Potential species-specific effects. Toxins. 2018;**10**:244. DOI: 10.3390/toxins10060244

[61] Scott PM. Recent research on fumonisins: A review. Food Additives and Contaminants. 2012;**29**(2):242-248. DOI: 10.1080/19440049.2010.546000

[62] Gadzala-Kopciuch R, Cendrowski K, Cesarz A, Kielbasa P, Buszewski B. Determination of zearalenone and its metabolites in endometrial cancer by coupled separation techniques. Analytical and Bioanalytical Chemistry. 2011;**401**(7):2069-2078

[63] Abia WA, Warth B, Sulyok M, Krska R, Tchana AN, Njobeh PB, et al. Determination of multi-mycotoxin occurrence in cereals, nuts and their products in Cameroon by liquid chromatography tandem mass spectrometry (LC-MS/MS). Food Control. 2013;**31**:438-453

[64] Kowalska K, Habrowska-Górczynska DE, Piastowska-Ciesielska AW. Zearalenone as an endocrine disruptor in humans. Environmental Toxicology and Pharmacology. 2016;**48**:141-149

[65] Tralamazza SM, Bemvenuti RH, Zorzet P, Garcia FS, Corrêa B. Fungal diversity and natural occurrence of deoxynivalenol and zearalenone in freshly harvested wheat grains from Brazil. Food Chemistry. 2016;**196**:445-450

[66] EFSA. Scientific opinion on the risks for human and animal health related to the presence of modified forms of certain mycotoxins in food and feed. EFSA Journal. 2014;**12**(12):3916

[67] Chatopadhyay P, Pandey A, Chaurasia AK, Upadhyay A, Karmakar S, Singh L. Hepatic hyperplasia and damages induces by zearalenone *Fusarium* mycotoxins in BALB/c mice. Arquivos de Gastroenterologia. 2012;**49**(1):77-81

[68] Sun LH, Lei MY, Zhang NY, Gao X, Li C, Krumm CS, et al. Individual and combined cytotoxic effects of aflatoxin B1, zearalenone, deoxynivalenol and fumonisin B1 on BRL 3A rat liver cells. Toxicon. 2015;**95**:6-12

[69] Rai A, Dixit S, Singh SP, Gautam NK, Das M, Tripathi A. Presence of zearalenone in cereal grains and its exposure risk assessment in Indian population. Journal of Food Science. 2018;**83**(12):1-8

[70] Chilaka CA, Boevre MD, Atanda OO, Saeger SD. The status of Fusarium Mycotoxins in sub-Saharan Africa: A review of emerging trends and post-harvest mitigation strategies towards. Food Control. 2017;**9**(1):19. DOI: 10.3390/toxins9010019

[71] Krska R, Baumgartner S, Josephs R. The state-of-the-art in the analysis of type-a and -B trichothecene mycotoxins in cereals. Fresenius' Journal of Analytical Chemistry. 2001;**371**:85-299

[72] Villafana RT, Ramdass AC, Rampersad SN. Selection of *Fusarium* trichothecene toxin genes for molecular detection depends on TRI gene cluster organization and gene function. Toxins. 2019;**11**:36. DOI: 10.3390/toxins11010036

[73] Vogelgsang S, Beyer M, Pasquali M, Jenny E, Musa T, Bucheli TD, et al. An eight-year survey of wheat shows distinctive effects of cropping factors on different *Fusarium* species and associated mycotoxins. European Journal of Agronomy. 2019;**105**:62-77

[74] Vogelgsang S, Sulyok M, Hecker A, Jenny E, Krska R, Schuhmacher R, et al. Toxigenicity and pathogenicity of *Fusarium poae* and *Fusarium avenaceum* on wheat. European Journal of Plant Pathology. 2008;**122**:265-276

[75] Sudakin DL. Trichothecenes in the environment: Relevance to human health. Toxicology Letters. 2003;**143**:97-107

[76] Pestka JJ. Deoxynivalenol: Toxicity, mechanisms and animal health risks. Animal Feed Science and Technology. 2007;**137**:283-298

[77] Dean R, Van Kan JA, Pretorius ZA, Hammond-Kosack KE, Di Pietro A, Spanu PD, et al. The top 10 fungal pathogens in molecular plant pathology. Molecular Plant Pathology. 2012;**13**:414-430

[78] Lowe R, Jubault M, Canning G, Urban M, Hammond-Kosack KE. The induction of mycotoxins by trichothecene producing *Fusarium* species. In: Plant Fungal Pathogens. Berlin, Germany: Springer; 2012. pp. 439-455

[79] Gautam P, Dill-Macky R. Type I host resistance and Trichothecene accumulation in *Fusarium*-infected wheat heads. American Journal of Agricultural and Animal Sciences. 2011;**6**(2):231-241

[80] Lindblad M, Gidlund A, Sulyok M, Börjesson C, Krska R, Olsen M, et al. Deoxynivalenol and other selected *Fusarium* toxins in Swedish wheat — Occurrence and correlation to specific *Fusarium* species. International Journal of Food Microbiology. 2013;**67**(2): 284-291

[81] Nagashima H. Toxicity of trichothecene mycotoxin nivalenol in human leukemia cell line HL60. JSM Mycotoxins. 2015;**65**(1):11-17

[82] Bryła M, Ksieniewicz-Wozniak E, Waskiewicz A, Szymczyk K, Jedrzejczak R. Natural occurrence of nivalenol, deoxynivalenol, and deoxynivalenol-3-glucoside in polish winter wheat. Toxins. 2018;**10**:81. DOI: 10.3390/toxins10020081

[83] Bennett JW, Klich M. Mycotoxins. Journal of Clinical Microbiology. 2003;**16**(3):497-516. DOI: 10.1128/cmr.16.3.497-516.2003

[84] Rocha O, Ansari K, Doohan FM. Effects of trichothecene mycotoxins on eukaryotic cells: A review. Food Additives and Contaminants. 2005;**22**:369-378

[85] Maresca M. From the gut to the brain: Journey and pathophysiological effects of the food-associated Trichothecene Mycotoxin Deoxynivalenol. Toxins. 2013;**5**:784-820. DOI: 10.3390/toxins5040784

[86] Rotter BA, Prelusky DB, Pestka JJ. Toxicology of deoxynivalenol (vomitoxin). Journal of Toxicology and Environmental Health. 1996;**48**(1):1-34

[87] Barkai-G R, Paster N. Mouldy fruits and vegetables as a source of mycotoxins: Part 1. World Mycotoxin Journal. 2008;**1**(2):147-159

[88] Rose LJ, Okoth S, Flett BC, van Rensburg BJ, Viljoen A. Preharvest Management Strategies and their Impact on Mycotoxigenic Fungi and Associated Mycotoxins. Rejika: IntechOpen; 2018. DOI: 10.5772/intechopn.76808

[89] Gil L, Ruiz P, Font G, Manyes L. An overview of the applications of hazards analysis and critical control point (HACCP) system to mycotoxins. Revista de Toxicología. 2016;**33**:50-55

[90] Bankole S, Schollenberger M, Drochner W. Mycotoxins in food systems in sub Saharan Africa: A review. Mycotoxin Research. 2006;**22**:163-169. DOI: 10.1007/BF02959270

[91] Ghosh MK, Chhabra A, Atreja PP, Chopra RC. Effect of treating with propionic acid, sodium bisulfite and sodium hydroxide on the biosynthesis of aflatoxin on groundnut cake. Animal Feed Science and Technology. 1996;**60**:43-49. DOI: 10.1016/0377-8401(95)00923-X

[92] Pascale M, Haidukowski M, Lattanzio VM, Silvestri M, Ranieri R, Visconti AJ. Distribution of T-2 and HT-2 toxins in milling fractions of durum wheat. Journal of Food Protection. 2011;**74**(10):1700-1707

[93] Scudamore KA, Patel S. The fate of deoxynivalenol and fumonisins in wheat and maize during commercial breakfast cereal production. World Mycotoxin Journal. 2008;**1**:437-448. DOI: 10.3920/WMJ2008.1059

[94] Scudamore KA, Patel S. Survey for aflatoxins, ochratoxin A, zearalenone and fumonisins in maize imported into the United Kingdom. Food Additives and Contaminants. 2000;**17**(5):407-416

[95] Park DL. Effect of processing on aflatoxin. In: DeVries JW, Trucksess MW, Jackson LS, editors. Mycotoxins and Food Safety. Boston, MA, USA: Springer; 2002. pp. 173-179

[96] Halasz A, Lasztity R, Abonyi T, Bata A. Decontamination of mycotoxin containing food and feed by biodegradation. Food Reviews International. 2009;**25**:284-298

[97] Ryu D, Jackson LS, Bullerman LB. Effects of processing on zearalenone. In: DeVries JW, Trucksess MW, Jackson LS, editors. Mycotoxins and Food Safety. Boston, MA, USA: Springer; 2002. pp. 205-216

[98] Kolosova A, Stroka J. Substances for reduction of the contamination of feed by mycotoxins: A review. World Mycotoxin Journal. 2011;**4**:225-256

[99] Adebiyi JA, Kayitesi E, Adebo OA, Changwa R, Njobeh PB. Food fermentation and mycotoxin detoxification: An African perspective. Food Control. 2019;**106**:106731. DOI: 10.1016/j.foodcont.2019.106731

[100] Hazard Analysis Critical Control Point (HACCP) 01/29/2018 U.S. Department of Health and Human Services, U.S. Food and Drug Administration. Available from: https://www.fda.gov/food/guidanceregulation/haccp/ [Accessed: 26 April 2019]

[101] Pineiro M, Nagler M, Coker R, Nicolaides L, Wareing P, et al. Manual on the Application of the HACCP System in Mycotoxin Prevention and Control. Rome: Food and Agriculture Organization of the United Nations; 2001. Available from: http://www.fao.org/3/a-y1390e.pdf [Accessed: 26 April 2019]

[102] Mycotoxins in Australian maize production: How to reduce the risk, National Research Centre for Environmental Toxicology (EnTox), University of Queensland; University of Sydney; Queensland Department of Primary Industries & Fisheries; NSW Department of Primary Industries; and the Grains Research & Development Corporation. Available from: http://www.maizeaustralia.com.au/mycotoxin_files/Mycotoxins%20in%20Australian%20maize%20production%20how%20to%20reduce%20the%20risk.pdf [Accessed: 26 April 2019]

[103] Training Modules on General Food Safety Plans for the Food Industry. APEC Secretariat, Michigan State University and The World Bank Group. 2012. Available from: http://fscf-ptin.apec.org/docs/APEC%20Food%20Safety%20Modules%202012/English%20Modules%20PDF/SCM_01_Introduction_6-2012-English.pdf [Accessed: 26 April 2019]

[104] Rychlik M, Humpf HU, Marko D, Dänicke S, Mally A, Berthiller F, et al. Proposal of a comprehensive definition of modified and other forms of mycotoxins including "masked" mycotoxins. Mycotoxin Research. 2014;**30**(4):197-205. DOI: 10.1007/s12550-014-0203-5

Chapter 5

Fusarium graminearum Species Complex and Trichothecene Genotype

Jianhua Wang, Zhiyong Zhao, Xianli Yang, Junhua Yang, Andong Gong, Jingya Zhang, Lei Chen and Changyan Zhou

Abstract

The fungal phytopathogen in *Fusarium* species can cause Fusarium head blight of wheat, barley, oats, and other small cereal grain crops worldwide. Most importantly, these fungi can produce different kinds of mycoxins, and they are harmful to humans and animal health. FAO reported that approximately 25% of the world's grains were contaminated by mycotoxins annually. This chapter will focus on several topics as below: (1) composition of *Fusarium graminearum* species complex; (2) genotype determination of *Fusarium graminearum* species complex strains from different hosts and their population structure changes; (3) genetic approaches to genotype determination in type B-trichothecene producing *Fusaria* fungi; and (4) some newly identified trichothecene mycotoxins, their toxicity, and distribution of the producers.

Keywords: *Fusarium graminearum* species complex, trichothecene, Fusarium mycotoxin, trichothecene genotype

1. Introduction

The fungal phytopathogen in *Fusarium graminearum* species complex (FGSC) are the primary etiological agent of Fusarium head blight (FHB) of wheat, barley, oats, and other small cereal grain crops worldwide. Besides, the Gibberella ear rot (GER) caused by FGSC and the related species *F. verticillioides* is one of the most devastating diseases on maize. FHB and GER are economically devastating plant disease that greatly limits grain yield and quality. Warm and humid weather conditions at the flowering stage are conducive to disease development. During the 1990s, economic losses in cereals (wheat and barley) caused by *Fusarium* were estimated at close to US $3 billion (US $2.5 billion in wheat and US $400 million in barley) and US $520 million (US $220 million in wheat and US $300 million in barley) in the United States and Canada, respectively [1]. It was reported that due to the changes in climatic conditions and in agricultural practices, outbreaks of FHB have occurred more frequent and serious in China. From 2008 to 2015, serious yield loss of wheat caused by FHB was occurred in more than 5 million ha each year.

In addition, infested grain is often contaminated with Fusarium toxins which are harmful to human and animal health and pose a serious threat to food or feed safety. FHB and GER are among the most destructive and economically important

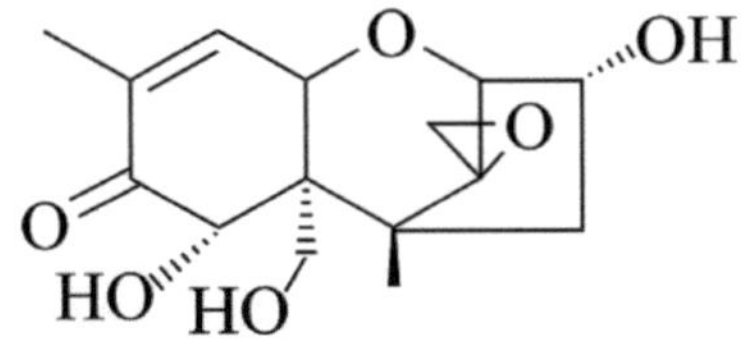

Figure 1.
Chemical structure of Deoxynivalenol (DON).

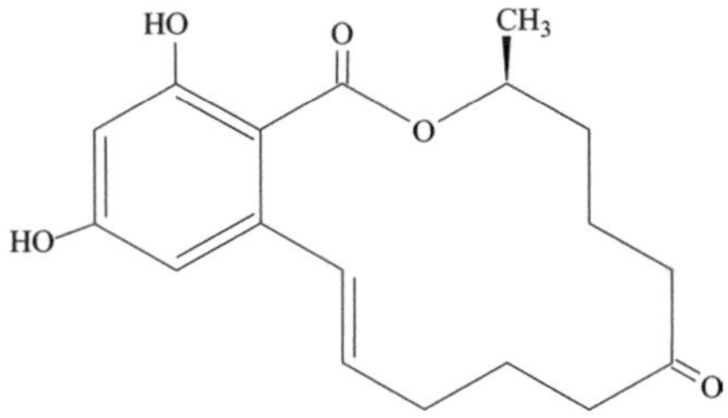

Figure 2.
Chemical structure of Zearalenone (ZEN).

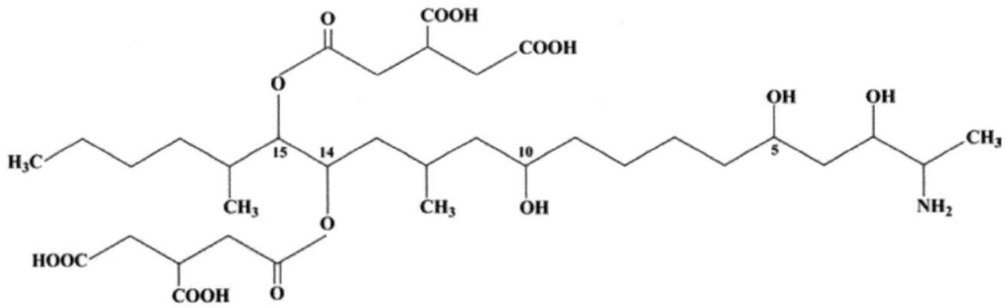

Figure 3.
Chemical structure of Fumonisin B_1 (FB_1).

diseases through the world. A survey made by the journal *Molecular Plant Pathology* from the international community, and resulted in the generation of a top 10 fungal plant pathogen list with FGSC in fourth place [2].

Up to now, more than 70 *Fusarium* species have been identified within the *Fusaria* genus. FGSC, *F. verticillioides*, *F. culmorum*, *F. oxysporum*, *F. solani*, *F. proliferatum*, *F. poae*, *F. equiseti*, and *F. fujikuroi* are the most commonly isolated species worldwide on wheat, maize and other plants. The most important thing is that, many different kinds of mycotoxins can be produced by these molds, such as deoxynivalenol (DON, **Figure 1**), zearalenone (ZEN, **Figure 2**), and fumonisin B_1 (FB_1, **Figure 3**) are the most prevalent Fusarium mycotoxins in cereal grains and they are very important in food and feed safety. It is clear now that one mold species may produce many different kinds of mycotoxins, and the same mycotoxin may be produced by several species. For example, FGSC can produce trichothecene and zearalenone, while trichothecene can be produced by FGSC, *F. culmorum*, *F. poae*, and *F. equiseti*. This chapter mainly focused on the FGSC and summarized the genetic methods used for trichothecene genotype determination of the strains.

2. Composition and identification of FGSC strains

Prior to 2000, due to the failure of morphological species recognition to accurately assess species limits for the FGSC, the species complex were considered a single cosmopolitan species. Applying the genealogical concordance phylogenetic species recognition (GCPSR), FGSC was first divided into seven phylogenetic lineages in 2000 [3]. Phylogenetic analyses of multilocus genotyping (MLGT) of DNA sequences from portions of 13 housing keeping genes, combined with GCPSR and molecular marker technologies, it revealed that this morphospecies comprises at least 16 biogeographically structured, phylogenetically distinct species. After that the species designation *Fusarium graminearum* have been *sensu stricto* in some conditions. Up to now, 15 of the 16 species have been formally described, including *F. acaciae-mearnsii*, *F. aethiopicum*, *F. asiaticum*, *F. austroamericanum*, *F. boothii*, *F. brasilicum*, *F. cortaderiae*, *F. gerlachii*, *F. graminearum sensu stricto*, *F. louisianense*, *F. meridionale*, *F. mesoamericanum*, *F. nepalense*, *F. ussurianum*, *F. vorosii*, and one additional species was informally recognized based on genealogical exclusivity and conidial morphology on SNA [4].

Proper species identification is critical to research aimed at improving disease and mycoxins control programs. However, it is difficult to discriminate the FGSC strains accurately by morphological characters. A partial region of the translation elongation factor 1 alpha gene (*TEF-1α*) was widely used for molecular identification of *Fusarium* genus. Some specific databases were created for *Fusarium* DNA sequence alignment analysis. For example, similarity searches of the obtained sequences can be performed with the Pairwise DNA alignments network service of the Fusarium MLST database (http://www.westerdijkinstitute.nl/fusarium/), Basic Local Alignment Search Tool (BLAST) network service of the Fusarium ID database (http://www.fusariumdb.org/index.php), and NCBI nucleotide database.

3. Mycotoxins produced by FGSC

In addition to yield reduction, the FGSC fungi are also of concern because they can produce different kinds of mycotoxins, e.g. zearalenone (**Figure 2**) and trichothecenes (**Figures 4** and **5**) in infested grains. Mycotoxin contamination can occur in both unprocessed and processed grains, representing a risk for human and animal health. Deleterious health effects caused by different mycotoxins include nephropathy, infertility, cancer or death [5].

Up to now, more than 200 trichothecenes have been identified [6]. Due to the chemical structure diverse, trichothecenes are divided into four types, namely type A (have a single bond at carbon atom 8, C-8), e.g. T-2 toxin (**Figure 4**), type B (have a keto at C-8), type C (have an epoxide at C-7, 8), and type D (have a macrocyclic ring between C-4 and C-15). All trichothecenes share a common tricyclic 12, 13-epoxytrichothec-9-ene, and they are derived from the isoprenoid intermediate farnesyl pyrophosphate via a series of biochemical reactions in *Fusarium*.

Among these mycotoxins, type B trichothecenes (**Figure 5**) are the most common detected in cereal grains and their related products. They are distinguished from type A by the presence of a keto function at C-8, and include deoxynivalenol (DON) and its acetylated forms 3-acetyldeoxynivalenol (3-ADON) and 15-acetyldeoxynivalenol (15-ADON), as well as nivalenol (NIV) and its acetylated form 4-acetylnivalenol (4-ANIV). DON is characterized by the absence of a hydroxyl

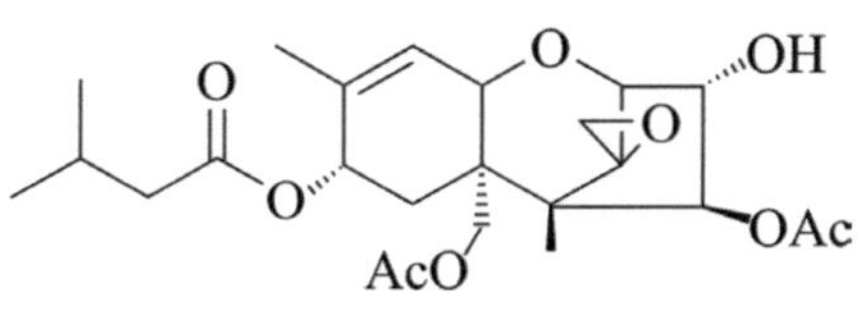

Figure 4.
Chemical structure of T-2 toxin.

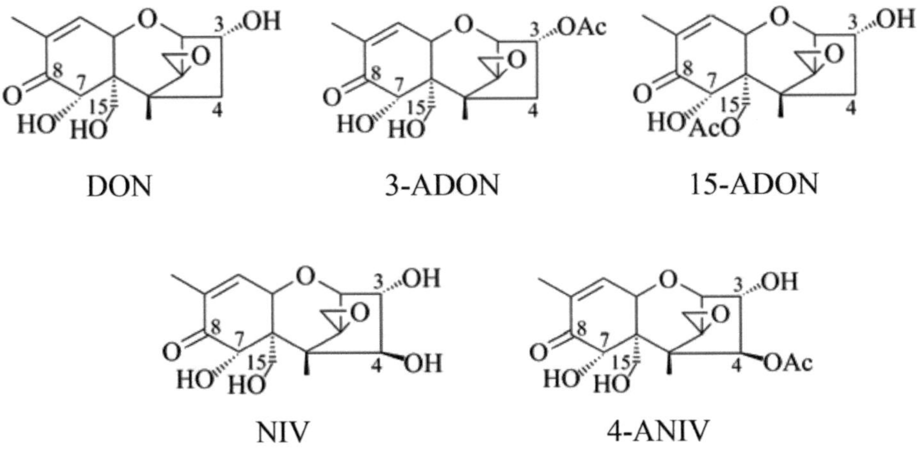

Figure 5.
Chemical structures of deoxynivalenol (DON), 3-acetyl deoxynivalenol (3-ADON), 15-acetyl deoxynivalenol (15-ADON), nivalenol (NIV), and 4-acetyl nivalenol (4-ANIV).

function at C-4, whereas NIV is characterized by the presence of a hydroxyl function at C-4. 3-ADON and 15-ADON are the acetylated forms of DON at C-3 and C-15, respectively. Meanwhile, NIV and 4-ANIV can be differed by the absence (NIV) and presence (4-ANIV) of an acetyl function at C-4 (**Figure 5**).

Type B trichothecenes are of the greatest concern in wheat and barley-growing regions worldwide, because they can represent a major threat to food and feed safety. These toxins are potent inhibitors of protein synthesis and are responsible for neurologic, gastrointestinal, immune function and other disorders. Although type B trichothecenes differ only slightly from each other in terms of the pattern and position of acetylation or hydroxylation, these changes can greatly affect the toxicity and the activity of these chemical compounds [7]. For example, DON is associated with feed refusal, vomiting and suppressed immune functions, and NIV is more toxic to humans and domestic animals than is DON. Determination of these trichothecene variations are important because the toxicity of DON and NIV may vary according to the eukaryotic organism affected. Minervini et al. [8] found that NIV was approximately four times more toxic than DON to human cells. Conversely, DON is 10 to 24 times more toxic to plant cells than NIV [9].

Type B trichothecenes are mainly produced by FGSC. Due to the ability of FGSC strains that producing different kinds of trichothecenes, three strain-specific trichothecene genotypes (chemotypes) were identified in FGSC: the 3-ADON genotype produces DON and 3-ADON, the 15-ADON genotype produces DON and 15-ADON, and the NIV genotype produces NIV and its acetylated derivatives 4-ANIV [10].

4. Mycotoxins act as virulence on certain hosts

Evidence is presented to show that trichothecene toxins act as virulence factors on certain hosts. Strains carrying a disrupted trichodiene synthase gene *Tri5* do not produce trichothecenes or their biosynthetic intermediates [11]. Disruption of the trichodiene synthase which catalyzes the first step in trichothecene biosynthesis in a 15-ADON producer GZ3639 exhibited reduced virulence on seedlings and heads of wheat, but wild-type virulence on seedlings of maize. The results indicated that trichothecene production contributes to the virulence of FGSC on wheat [12]. Eudes et al. [13] confirmed earlier findings that trichothecenes are a principal determinant of *F. graminearum* aggressiveness on wheat cultivars. More recently, Maier et al. [14] investigated the involvement of trichothecenes in the virulence of the pathogen by disrupting the *Tri5* gene encoding the first committed enzyme trichodiene synthase in FGSC strains with 3-ADON, 15-ADON, and NIV genotype, respectively. The results demonstrated that disruption mutants can cause disease symptoms on the inoculated spikelet but the symptoms cannot spread into other spikelets on wheat. And on maize, mutants derived from the NIV genotype strain caused less disease than their progenitor strain, while no significant difference compared to the wild-type strains were observed on barley [14]. Trichothecenes are therefore also potent phytotoxins and act as virulence factors of pathogenic fungi thus facilitate tissue colonization on sensitive host plants, e.g. of DON produced by *F. graminearum* in wheat [15].

Host preference was identified among the FGSC on wheat, maize, barley, and rice in certain regions. Several studies suggested that maize played a significant role for the presence of the NIV genotype for FGSC. NIV producers were found to be more aggressive toward maize compared to DON producers [16]. This can be due to the fact that NIV is a virulence factor useful for maize colonization [14], and therefore the plant probably represents an ecological niche for hosting the NIV genotype strains. A high proportion of NIV producers of FGSC on maize were observed in China by our group based on a collection from 59 districts in 19 provinces throughout China, and NIV producers preferentially to maize [17].

5. Distribution and population changes of FGSC

The species composition and genotype prevalence of FGSC vary widely in different regions, which reflecting the level of risk factors in feed/food safety. Investigations on *Fusarium* species isolated from wheat, barley, and maize crops have been reported in the last two decades. Dynamic changes of species composition and chemotype proportion have been found in different agricultural ecosystems worldwide. Prior to 2000, strains from the United States and Canada were almost exclusively 15-ADON producers, while they have been increasingly replaced by the 3-ADON producers in some major wheat-growing regions, e.g. the frequency of the 3-ADON genotype in western Canada increased more than 14-fold between 1998 and 2004 [18]. Also Schmale et al. [19] analyzed the trichothecene genotypes of *Gibberella zeae* collected from winter wheat fields in the eastern parts of the US. They revealed an increasing gradient in 3-ADON distribution from south to north and closer to Canada. In some regions, the *F. graminearum* 15-ADON chemotype is being replaced by the 3-ADON chemotype. The epidemiology data indicated that 3-ADON chemotype dominates in northern Europe, while 15-ADON chemotype dominates in North America, central Europe and southern Russia and some parts of Asia.

The composition of FGSC population appears to be host and location dependent. The results by Zhang et al. [20] and Shen et al. [21] indicated that *F. asiaticum* was the predominated in wheat. Among the 97 FGSC assayed from rice (30 strains), maize (33 strains), and wheat (34 strains) by Qiu and Shi [22], 73 strains were identified as *F. asiaticum* and 63 of them were collected from rice or wheat. The remaining 24 strains belonged to *F. graminearum sensu stricto* and 23 of them were isolated from maize, only 1 strain was collected from wheat. Similarly, FGSC strains were isolated from GER samples in South Korea with *F. graminearum sensu stricto* to be the dominant species which account for 75% of the FGSC [23], while *F. asiaticum* was the dominant species (78.5%) on Korean rice and followed by *F. graminearum sensu stricto* [24].

The distribution of FGSC may correlate with annual temperature. Qu et al. [25] reported that temperature affected the geographic distribution of *F. graminearum sensu stricto* and *F. asiaticum* on wheat spikes in China. A comprehensive study on FGSC from wheat was conducted by Zhang et al. [20]. They found that the geographic distribution of FGSC associated with the annual average temperature. The cooler temperatures (annual average temperature ≤15°C) appear to favor *F. graminearum sensu stricto*, while the warmer regions (annual average temperature ≥15°C) appear to favor *F. asiaticum*. A hypothesis was made that the distribution of FGSC members are climate dependent [20].

F. graminearum sensu stricto with the 15-ADON genotype and *F. asiaticum* with either the NIV or the 3-ADON genotype were the dominant causal agents on wheat, and the two species dominated the northern and southern regions of China, respectively, which is consistent with earlier studies [20, 26, 27].

However, more recently the study by Zhang et al. [28] indicated that temperature may not be the only factor in the distribution of FGSC and that other, yet unknown factors affected their distribution. To explain genotype distribution in different geographic areas, hypotheses based on grain seed shipment, international trade, long-distance spore transportation, and environmental favorable conditions were proposed.

6. FGSC fitness vary

Phylogenetic analyses of trichothecene gene cluster demonstrated that genotype polymorphism is trans-specific and have been maintained by balancing selection on the ancestral pathogens, and genotype differences may have a significant impact on pathogen fitness [29].

The FGSC strains with different genotype showed different fitness to the ecological environment, such as the hosts, temperature, rotation, and so on. 3-ADON producer was more aggressive than 15-ADON population in susceptible wheat, and also the 3-ADON isolates exhibit a higher DON production than the 15-ADON isolates. Similar conclusions were made by Zhang et al. [28] that *F. asiaticum* strains with 3-ADON chemotype revealed significant advantages over the strains that produce NIV in pathogenicity, growth rate, trichothecene accumulation, etc. Their data also indicated that the growth of rice may be a key factor for the presence of *F. asiaticum* [28]. Liu et al. [30] compared the fitness of three chemotype *Fusarium* strains, and found that 15ADON producers had the advantage in perithecia formation and ascospore release, whereas more DON were produced by the 3-ADON chemotypes. Qiu and Shi [22] estimated the effect of rice or maize as former crops on mycotoxin accumulation in wheat grains, and they concluded that rice-wheat rotation favors DON accumulation.

Changes in DON chemotypes distribution were reported for FGSC from Canada, USA, and Northern Europe. Recently, Nicolli et al. [31] assessed a range of fitness-related traits (perithecia formation, mycelial growth, sporulation and germination, pathogenicity, and sensitivity to tebuconazole) with 30 strains representatives of 3ADON-, 15ADON-, and NIV-producers. The pathogenicity assay results indicated that strains with the DON chemotypes were generally more aggressive than the NIV ones [31].

Phenotypic analyses indicated that *F. asiaticum* with a 3-ADON genotype revealed significant advantages over *F. asiaticum* that produce NIV in pathogenicity, growth rate, and trichothecene mycotoxin accumulation. It shall be noted that a biased gene flow from 3-ADON to NIV producers was identified in *F. asiaticum* from wheat in China [28].

FGSC from wheat-maize rotation regions on wheat spikes and maize stalks in Henan province, China, was determined by Hao et al. [32], and significant differences were found in the frequencies of *F. graminearum sensu stricto* and *F. asiaticum* species within the hosts with *F. graminearum sensu stricto* to be the dominant. Genotype analysis revealed that 15-ADON producers represented 92.7 and 98.5% of isolates from wheat and maize, respectively. The three genotypes may affect species distribution or population ecology because these mycotoxins are differing in toxicity and bioactivity [7, 29, 33].

7. Genetic genotype determination of FGSC

Traditionally, chemotyping of FGSC strains has been carried out using gas chromatography/mass spectroscopy. This method can be time-consuming and expensive. The genome sequences of several FGSC strains have been published. The trichothecene core gene cluster nucleotide sequences of many strains representatives 3-ADON, 15-ADON, and NIV genotypes have also been deposited in the GenBank. The availability of this information makes it possible to reveal the structural features and allowed selection of several primer sets used successfully in PCR experiments for the molecular characterization of the various chemotypes. Molecular genetic assays allow for high throughput screening of large numbers of field isolates.

Lee et al. [34] sequenced the gene cluster for trichothecene biosynthesis from a 15-ADON producer (strain H-11) and a NIV producer (strain 88-1), and sequence polymorphisms within the *Tri7* open reading frame was found between the two strains. Alignment analysis suggesting that the *Tri7* gene of H-11 carried several mutations and an insertion compared to the *Tri7* gene from 88-1, and based on the sequence difference a PCR-based diagnostic method for differentiating DON and NIV producers by polyacrylamide gel electrophoresis was developed.

Lee et al. [35] subsequently sequenced the *Tri13* homolog from DON (strain H-11) and NIV producers (strain 88-1) and found that the gene differs drastically between the two producers, suggesting that the *Tri13* gene could be used for genetic genotype distinction for DON and NIV producers [35, 36]. They further confirmed the roles of the *Tri7* and *Tri13* genes in trichothecene production, and the results suggested that both the *Tri7* and *Tri13* genes are nonfunctional in DON producers [35].

The PCR assays to *Tri7* and *Tri13* genes developed by Lee et al. [34, 35] allowed clear differentiation between DON and NIV genotypes. However, they could not be used to further classify the DON-producing isolates to 3-ADON or 15-ADON producer. Ward et al. [29] examined a 19-kb region of the trichothecene gene cluster that sequenced in 39 strains representing 3-ADON, 15-ADON, and NIV genotypes. They found that Tri-cluster haplotypes group according to genotype rather than by species indicated that 3-ADON, 15-ADON, and NIV genotypes each have a single evolutionary origin. Reciprocally monophyletic groups, corresponding to each of 3-ADON, 15-ADON, and NIV genotypes, were strongly supported in *Tri3*, *Tri11*, and *Tri12* genes trees. Two sets of primers specific to the individual genotypes were designed from *Tri3* and *Tri12* genes. The genotype-specific PCR tests developed by Ward et al. [29] provide a rapid and direct genetic method for distinguishing among 3-ADON, 15-ADON, and NIV producer, this is the first report differentiated these three genotype strains by a PCR method.

The work by Lee et al. [34, 35] and Brown et al. [37] indicated that the genes *Tri13* and *Tri7* from trichothecene biosynthetic cluster are responsible for conversion of DON to NIV (*Tri13* gene) and the *Tri7* gene product modifies NIV by acetylation of C-4 atom hydroxyl to produce 4-ANIV. Based on these results sets of positive-negative PCR assays to *Tri7* and *Tri13* genes for trichothecene determination of FGSC were developed by Chandler et al. [38], and the assays can accurately indicate a DON or NIV genotype in FGSC, *F. culmorum* and *F. cerealis*. The assays were successfully used to screen isolates from different countries and the genotype-specific assays were able to detect and characterize a wider range of species and haplotypes than previous methods.

By comparing the published sequences for *Tri13* gene from known DON- and NIV- producers, Waalwijk et al. [39] designed a primer pair to discriminate the two genotypes which generated a 234 bp fragment in DON-producers and a fragment of 415 bp in NIV-producers. The *Tri13* primer pair was capable and robust to determine the genotype of strains from *F. culmorum*.

Based on information reported and deposited by Ward et al. [29], three primer sets were designed to the *Tri3* gene by Jennings et al. [40] to allow further differentiation of the DON genotype into either 3-ADON or 15-ADON. Each isolate produces a PCR product with only one of these primer sets but not the other two from *F. culmorum* and FGSC strains [40, 41].

Li et al. [42] found that the intergenic sequences between *Tri5* and *Tri6* genes appear to be mycotoxin genotype-specific, and based on the sequence length polymorphism a generic PCR assay was developed to detect a 300 bp fragment of DON-genotype strains and a 360 bp fragment of NIV-genotypes from FGSC.

Based on the sequences of FGSC described by Lee et al. [34] and Ward et al. [29], a series of PCR assays have been designed to *Tri3* and *Tri7* by Quarta et al. [43], in order to permit specific detection of 3-ADON, 15-ADON, and NIV genotypes, respectively. These primers were subjected to a multiplex PCR assay for the identification of the different genotypes of *Fusarium* strains combined with the primer pair derived from the *Tri5* gene by Bakan et al. [44]. The multiplex PCR was validated on FGSC, *F. cerealis*, *F. culmorum* strains from different European countries, and successfully used to identify the genotype of the *Fusarium* strain contaminating wheat kernels [43, 45].

The possibility to distinguish by a singleplex PCR 3-ADON, 15-ADON, and NIV genotypes was not yet resolved until very recently. Wang et al. [46] developed a *Tri13* based PCR assay and successfully identified the 3-ADON, 15-ADON, and NIV genotypes in FGSC from Asia, Europe, and America. Using the primer pair, specific amplification products of 644, 583, and 859 bp were obtained from isolates producing 3-ADON, 15-ADON, and NIV, respectively. All three types of PCR fragments had different molecular sizes with a smallest difference of 61 bp can be directly differentiated on an agarose gel. The method should be more reliable than other PCR-based assays that show the absence or presence of a PCR fragment since these assays may generate false-negative results. This is a rapid, reliable and cost-effective method for the determination of 3-ADON, 15-ADON, and NIV genotype strains in FGSC.

Recently Suzuki et al. [47] reported a multiplex PCR assay for simultaneous identification of the species and trichothecene genotypes for *F. graminearum sensu stricto* and *F. asiaticum* based on *Tri3* and *Tri6* genes. This approach proved successful for Japanese strains [47].

An alternative method based on *Tri11* polymorphism was developed by Zhang et al. [48] to differentiate 3-ADON, 15-ADON, and NIV genotypes of FGSC strains. Similarly, we presented another multiplex assay based on the single nucleotide polymorphism of *Tri11* gene between strains of different genotype [49]. The assay was also validated on plant material.

Recent work by Kulik [50] and Nielsen et al. [51] to detect and quantify FGSC genotypes in plants/grains were developed based on TaqMan probe set and SYBR green method with *Tri12* gene, respectively.

Due to the toxicological differences between DON and NIV, it is important to monitor the population and determine the chemotypes of strains present in any given geographic region. Mycotoxin producing capability of a certain strain could be established both through biochemical and molecular techniques.

Target gene	Primers	Sequences (5′ to 3′)	Fragment size (bp)	Chemotypes	References
Tri3	3CON	TGGCAAAGACTGGTTCAC	243	3-ADON	Ward et al. [29]
	3D3A	CGCATTGGCTAACACATG			
	3CON	TGGCAAAGACTGGTTCAC	610	15-ADON	
	3D15A	ACTGACCCAAGCTGCCATC			
	3CON	TGGCAAAGACTGGTTCAC	840	NIV	
	3NA	GTGCACAGAATATACGAGC			
	Tri303F	GATGGCCGCAAGTGGA	586	3-ADON	Jennings et al. [40, 41]
	Tri303R	GCCGGACTGCCCTATTG			
	Tri315F	CTCGCTGAAGTTGGAC GTAA	864	15-ADON	
	Tri315R	GTCTATGCTCTCAACG GACAAC			
	Tri3NivF	GGACGTGA(CG)TACT CTTGGCAA	549	NIV	
	Tri3NivR	CCCAG(AG)GCCTCTA AGAA(AG)GGB			
	Tri3F971	CATCATACTCGC TCTGCTG	708	15-ADON	Quarta et al. [43]
	Tri3R1679	TT(AG)TAGTTTGCATC ATT(AG)TAG			
	Tri3F1325	GCATTGGCTAACACATGA	354	3-ADON	
	Tri3R1679	TT(AG)TAGTTTGCA TCATT(AG)TAG			
	3D15AF	AACTGACCCAAGCTG CCATC	420	15-ADON (*F. asiaticum* and *F. graminearum ss*)	Suzuki et al. [47]

Target gene	Primers	Sequences (5′ to 3′)	Fragment size (bp)	Chemotypes	References
	3D15AR	CTTCTGTCCCTTCG AACGGA			
Tri5- *Tri6* intergenic	ToxP1	GCCGTGGGG(AG)TAA AAGTCAAA	300	DON	Li et al. [42]
region	ToxP2	TGACAAGTCCGGTC GCACTAGCA	360	NIV	
Tri6	6A3AF	CCAAGACTT(GT)GTT (AC)CCCGAA	1100	DON (*F. asiaticum*)	Suzuki et al. [47]
	6A3AR	GCAATCTTTAGAGTG CCGAC			
	6G3AF	T(AG)TCCCATCCCAT CAAGGCT	330	DON (*F. graminearum ss*)	
	6G3AR	AACAAGTGGTTCTT CGGAGT			
	6CNF	CAAGCAAATGCCC GTATCCC	660	NIV (*F. asiaticum*)	
	6ANR	CGCAACAATATCA ATGGCTGTGCTA			
Tri7	GzTri7/f1	GGCTTTACGACTC CTCAACAATGG	173–327	15-ADON	Lee et al. [34]
	GzTri7/r1	AGAGCCCTGCGAA AG(CT)ACTGGTGC	161	NIV	
	Tri7F	TGCGTGGCAATATC TTCTTCTA	458–535	DON	Chandler et al. [38]
	Tri7R	TGTGGAAGCCGCAGA	436	NIV	
	Tri7F	TGCGTGGCAATAT CTTCTTCTA	381–445	DON	

Target gene	Primers	Sequences (5′ to 3′)	Fragment size (bp)	Chemotypes	References
	Tri7DON	GTGCTAATATTGT GCTAATATTGTGC			
	Tri7F	TGCGTGGCAATAT CTTCTTCTA	465	NIV	
	Tri7NIV	GGTTCAAGTAAC GTTCGACAATAG			
	MinusTri7F	TGGATGAATGAC TTGAGTTGACA	483	3-ADON	
	MinusTri7R	AAAGCCTTCATT CACAGCC			
	Tri7F340	ATCGTGTACAAG GTTTAC G	625	NIV	Quarta et al. [43]
	Tri7R965	TTCAAGTAACGT TCGACAAT			
Tri11	3D11	GCAAGTCTGGC GAGGCC	342	3-ADON	Zhang et al. [48]
	11R	TCAAAGGCCAG AGCAACCC			
	15D11	AAGTATGGTCC AGTTGTCCGTATT	424	15-ADON	
	11R	TCAAAGGCCAG AGCAACCC			
	N11	CTTGTCAGGCGG CACAGTAG	643	NIV	
	11R	TCAAAGGCCAGA GCAACCC			

Target gene	Primers	Sequences (5′ to 3′)	Fragment size (bp)	Chemotypes	References
	Tri11-CON	GACTGCTCATGG AGACGCTG	334	3-ADON	Wang et al. [49]
	Tri11-3AcDON	TCCTCATGCTCG GTGGACTCG			
	Tri11-CON	GACTGCTCATGG AGACGCTG	279	15-ADON	
	Tri11-15AcDON	TGGTCCAGTTG TCCGTATT			
	Tri11-CON	GACTGCTCATG GAGACGCTG	497	NIV	
	Tri11-NIV	GTAGGTTCCAT TGCTTGTTC			
Tri12	12CON	CATGAGCATGG TGATGTC	410	3-ADON	Ward et al. [29]
	12-3F	CTTTGGCAAGC CCGTGCA			
	12CON	CATGAGCATGG TGATGTC	670	15-ADON	
	12-15F	TACAGCGGTCG CAACTTC			
	12CON	CATGAGCATGG TGATGTC	840	NIV	
	12NF	TCTCCTCGTTG TATCTGG			
Tri13	GzTri13/p1	AATACTA(CA)AAG(CT) CTAG(GT)ACGACGC	470	DON	Kim et al. [36]

Target gene	Primers	Sequences (5′ to 3′)	Fragment size (bp)	Chemotypes	References
	GzTri13/p2	GTG(AG)T(AG)TCCCA GGATCTGCGTGTC	760	NIV	
	Tri13F	TACGTGAAACAT TGTTGGC	234	DON	Waalwijk et al. [39]
	Tri13R	GGTGTCCCAGGA TCTGCG	415	NIV	
	Tri13F	CATCATGAGACTTGT (GT)C(AG)AGTTTGGG	282	DON	Chandler et al. [38]
	Tri13DONR	GCTAGATCGATT GTTGCATTGAG			
	Tri13NIVF	CCAAATCCGAA AACCGCAG	312	NIV	
	Tri13R	TTGAAAGCTCC AATGTCGTG			
	Tri13F	CATCATGAGACTTGT (GT)C(AG)AGTTTGGG	799	DON	
	Tri13R	TTGAAAGCTCC AATGTCGTG	1075	NIV	
	Tri13P1	CTC(CG)ACCGCATC GAAGA(CG)TCTC	583	15-ADON	Wang et al. [46]
	Tri13P2	GAA(CG)GTCGCA (AG)GACCTTGTTTC	644	3-ADON	
			859	NIV	

Table 1.
Primers designed for genetic genotyping of FGSC so far.

The biochemical approach involves the incubation and extraction of mycotoxins, the methods being complicated and time consuming. The molecular techniques are based on detection of specific gene by using specific primers. All these molecular methods developed for genotype analysis are based on nucleotide diversity of trichothecene synthesis genes. Chemotype characterization has been extensively used to characterize FGSC for their toxigenic potential [52]. The information about the genetic genotyping methods developed so far, such as targeted gene, primer name, primer sequence, and amplification fragment sizes are summarized in **Table 1**.

More effective and accuracy genetic methods are needed. We are doing genomic sequencing of FGSC strains with different trichothecene genotypes, and we believe some new molecular genetic methods will be developed based on the genomic data.

8. Newly identified trichothecene mycotoxins

In addition to the well characterized fungal mycotoxins, plant-derived mycotoxin metabolites, masked mycotoxins, have emerged as important co-contaminants in cereals [53, 54]. The most commonly detected masked mycotoxin conjugates are β-linked glucose-conjugates of trichothecenes, such as DON-3-glucoside (**Figure 6**). The possible hydrolysis of masked mycotoxins back to their toxic parents during mammalian digestion raises great concerns. Recently, a new series of type-A trichothecene, NX-toxins (**Figure 7**), produced by FGSC were characterized [5]. In vitro translation assays indicated that NX-3 can inhibit protein biosynthesis to almost the same extent as DON [5]. Comprehensive work on intestinal hydrolysis, absorption, metabolism, and toxicity of newly characterized mycotoxins need to be determined (**Figure 7**).

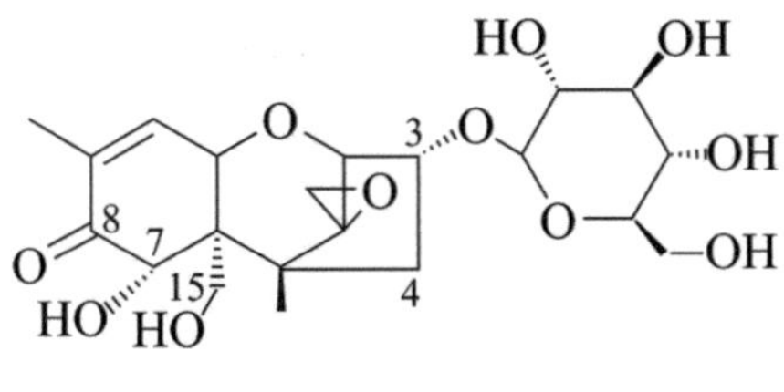

Figure 6.
Chemical structure of deoxynivalenol-3-glucoside (D3G).

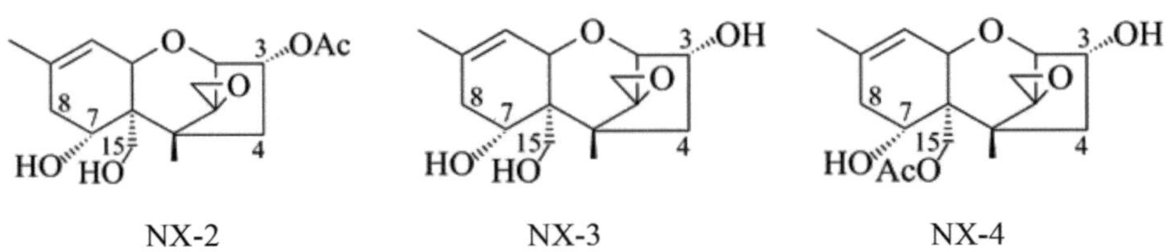

Figure 7.
Chemical structures of NX-2, NX-3, and NX-4.

9. Conclusion

The knowledge about the mycotoxins chemotypes could contribute to a better management of fungal infections and breeding of resistance, in order to obtain grains of better quality. The results will also contribute to improve our understanding of the ecology and epidemiology of FGSC members, which may be of value for improving models for assessing the risk or epidemics and mycotoxin production. Genetic genotyping has been proved to be a useful tool for predicting trichothecene type produced by FGSC, and future work on the more effective tools for genotype determination is needed. The discovery of novel toxic metabolites belonging to trichothecenes, such as NX-toxins is also suggesting that the prevalence, distribution, and genetic diversity of FGSC require continuous monitoring. Further research on the biosynthesis molecular mechanism of trichothecene, especially the novel mycotoxins is needed.

Acknowledgements

The authors acknowledged the financial support from National Natural Science Foundation of China (31871896, 31602124, and 31401598), Shanghai Agriculture Commission Basic Research Project (Grant number 2014 NO.7-3-7), and Shanghai Academy of Agricultural Sciences (Grant number 210ZJ1812).

Author details

Jianhua Wang[1*], Zhiyong Zhao[1], Xianli Yang[1], Junhua Yang[1], Andong Gong[2], Jingya Zhang[1], Lei Chen[1] and Changyan Zhou[1]

1 Institute for Agro-food Standards and Testing Technology, Laboratory of Quality and Safety Risk Assessment for Agro-products (Shanghai), Ministry of Agriculture, Shanghai Academy of Agricultural Sciences, Shanghai, P. R. China

2 College of Life and Science, Xinyang Normal University, Xinyang, P. R. China

*Address all correspondence to: jianhuawang163@163.com

References

[1] Windels CE. Economic and social impacts of Fusarium head blight: Changing farms and rural communities in the Northern Great Plains. Phytopathology. 2000;**90**(1):17-21

[2] Dean R, Van Kan JA, Pretorius ZA, Hammond-Kosack KE, Di Pietro A, Spanu PD, et al. The top 10 fungal pathogens in molecular plant pathology. Molecular Plant Pathology. 2012;**13**(4):414-430

[3] O'Donnell K, Kistler HC, Tacke BK, Casper HH. Gene genealogies reveal global phylogeographic structure and reproductive isolation among lineages of *Fusarium graminearum*, the fungus causing wheat scab. Proceedings of the National Academy of Sciences of the United States of America. 2000;**97**:7905-7910

[4] Aoki T, Ward TJ, Kistler HC, O'Donnell K. Systematics, phylogeny and trichothecene mycotoxin potential of Fusarium head blight cereal pathogens. Mycotoxins. 2012;**62**(2):91-102

[5] Varga E, Wiesenberger G, Hametner C, Ward TJ, Dong Y, Schöfbeck D, et al. New tricks of an old enemy: Isolates of *Fusarium graminearum* produce a type A trichothecene mycotoxin. Environmental Microbiology. 2015;**17**(8): 2588-2600

[6] Grove JF. The trichothecenes and their biosynthesis. Fortschritte der Chemie Organischer Naturstoffe. 2007;**88**:63-130

[7] Kimura M, Kaneko I, Komiyama M, Takatsuki A, Koshino H, Yoneyama K, et al. Trichothecene 3-O-acetyltransferase protects both the producing organism and transformed yeast from related mycotoxins. Journal of Biological Chemistry. 1998;**273**:1654-1661

[8] Minervini F, Fornelli F, Flynn KM. Toxicity and apoptosis induced by the mycotoxins nivalenol, deoxynivalenol and fumonisin B1 in a human erythroleukemia cell line. Toxicology In Vitro. 2004;**18**(1):21-28

[9] Shimada T, Otani M. Effects of Fusarium mycotoxins on the growth of shoots and roots at germination in some Japanese wheat cultivars. Cereal Research Communications. 1990;**18**:229-232

[10] Miller JD, Greenhalgh R, Wang YZ, Lu M. Trichothecene chemotypes of three *Fusarium* species. Mycologia. 1991;**83**(2):121-130

[11] Hohn TM, Desjardins AE. Isolation and gene disruption of the *Tox5*, gene encoding trichodiene synthase in *Gibberella pulicaris*. Molecular Plant-Microbe Interactions. 1992;**5**:249-256

[12] Proctor RH, Hohn TM, McCormick SP. Reduced virulence of *Gibberella zeae* caused by disruption of a trichothecene toxin biosynthetic gene. Molecular Plant-Microbe Interactions. 1995;**8**:593-601

[13] Eudes F, Comeau A, Rioux S, Collin J. Impact of trichothecenes on Fusarium head blight *Fusarium graminearum* development in spring wheat (*Triticum aestivum*). Canadian Journal of Plant Pathology. 2001;**23**:318-322

[14] Maier FJ, Miedaner T, Hadeler B, Felk A, Salomon S, Lemmens M, et al. Involvement of trichothecenes in fusarioses of wheat, barley and maize evaluated by gene disruption of the trichodiene synthase (*Tri5*) gene in three field isolates of different chemotype and virulence. Molecular Plant Pathology. 2006;**7**(6):449-461

[15] Jansen C, von Wettstein D, Schafer W, Kogel KH, Felk A, Maier FJ.

Infection patterns in barley and wheat spikes inoculated with wild-type and trichodiene synthase gene disrupted *Fusarium graminearum*. Proceedings of the National Academy of Sciences of the United States of America. 2005;**102**:16892-16897

[16] Carter JP, Rezanoor HN, Holden D, Desjardins AE, Plattner RD, Nicholson P. Variation in pathogenicity associated with the genetic diversity of *Fusarium graminearum*. European Journal of Plant Pathology. 2002;**108**:573-583

[17] Ndoye M, Zhang JB, Wang JH, Gong AD, Li HP, Qu B, et al. Nivalenol and 15-acetyldeoxynivalenol chemotypes of *Fusarium graminearum* clade species are prevalent on maize throughout China. Journal of Phytopathology. 2012;**160**:519-524

[18] Ward TJ, Clear RM, Rooney AP, O'Donnell K, Gaba D, Patrick S, et al. An adaptive evolutionary shift in Fusarium head blight pathogen populations is driving the rapid spread of more toxigenic *Fusarium graminearum* in North America. Fungal Genetics and Biology. 2008;**45**:473-484

[19] Schmale DG, Wood Jones AK, Cowger C, Bergstrom GC, Arellano C. Trichothecene genotypes of *Gibberella zeae* from winter wheat fields in the eastern USA. Plant Pathology. 2011;**60**:909-917

[20] Zhang JB, Li HP, Dang FJ, Qu B, Xu YB, Zhao CS, et al. Determination of the trichothecene mycotoxin chemotypes and associated geographical distribution and phylogenetic species of the *Fusarium graminearum* clade from China. Mycological Research. 2007;**111**:967-975

[21] Shen CM, Hu YC, Sun HY, Li W, Guo JH, Chen HG. Geographic distribution of trichothecene chemotypes of the *Fusarium graminearum* species complex in major winter wheat production areas of China. Plant Disease. 2012;**96**:1172-1178

[22] Qiu JB, Shi JR. Genetic relationships, carbendazim sensitivity and mycotoxin production of the *Fusarium graminearum* population from maize, wheat and rice in Eastern China. Toxins. 2014;**6**:2291-2309

[23] Lee J, Kim H, Jeon JJ, Kim HS, Zeller KA, Carter LL, et al. Population structure of and mycotoxin production by *Fusarium graminearum* from maize in South Korea. Applied and Environmental Microbiology. 2012;**78**:2161-2167

[24] Lee J, Chang IY, Kim H, Yun SH, Leslie JF, Lee YW. Genetic diversity and fitness of *Fusarium graminearum* populations from rice in Korea. Applied and Environmental Microbiology. 2009;**75**:3289-3295

[25] Qu B, Li HP, Zhang JB, Xu YB, Huang T, Wu AB, et al. Geographic distribution and genetic diversity of *Fusarium graminearum* and *F. asiaticum* on wheat spikes throughout China. Plant Pathology. 2008;**57**:15-24

[26] Gale LR, Chen LF, Hernick CA, Takamura K, Kistler HC. Population analysis of *Fusarium graminearum* from wheat fields in eastern China. Phytopathology. 2002;**92**:1315-1322

[27] Yang L, van der Lee T, Yang X, Yu D, Waalwijk C. *Fusarium* populations on Chinese barley show a dramatic gradient in mycotoxin profiles. Phytopathology. 2008;**98**:719-727

[28] Zhang H, Van der Lee T, Waalwijk C, Chen W, Xu J, Xu J, et al. Population analysis of the *Fusarium graminearum* species complex from wheat in China show a shift to more aggressive isolates. PLoS One. 2012;**7**(2):e31722

[29] Ward TJ, Bielawski JP, Kistler HC, Sullivan E, O'Donnell K. Ancestral polymorphism and adaptive evolution in the trichothecene mycotoxin gene cluster of phytopathogenic *Fusarium*. Proceedings of the National Academy of Sciences of the United States of America. 2002;**99**:9278-9283

[30] Liu YY, Sun HY, Li W, Xia YL, Deng YY, Zhang AX, et al. Fitness of three chemotypes of *Fusarium graminearum* species complex in major winter wheat-producing areas of China. PLoS One. 2017;**12**(3):e0174040

[31] Nicolli CP, Machado FJ, Spolti P, Del Ponte EM. Fitness traits of deoxynivalenol and nivalenol-producing *Fusarium graminearum* species complex strains from wheat. Plant Disease. 2018;**102**(7):1341-1347

[32] Hao JJ, Xie SN, Sun J, Yang GQ, Liu JZ, Xu F, et al. Analysis of *Fusarium graminearum* species complex from wheat–maize rotation regions in Henan (China). Plant Disease. 2017;**101**:720-725

[33] Alexander NJ, Hohn TM, McCormick SP. The *TRI11* gene of *Fusarium sporotrichioides* encodes a cytochrome P-450 monooxygenase required for C-15 hydroxylation in trichothecene biosynthesis. Applied and Environmental Microbiology. 1998;**64**:221-225

[34] Lee T, Oh DW, Kim HS, Lee J, Kim YH, Yun SH, et al. Identification of deoxynivalenol- and nivalenol-producing chemotypes of *Gibberella zeae* by using PCR. Applied and Environmental Microbiology. 2001;**67**:2966-2972

[35] Lee T, Han YK, Kim KH, Yun SH, Lee YW. *Tri13* and *tri7* determine deoxynivalenol- and nivalenol-producing chemotypes of *Gibberella zeae*. Applied and Environmental Microbiology. 2002;**68**:2148-2154

[36] Kim HS, Lee T, Dawlatana M, Yun SH, Lee YW. Polymorphism of trichothecene biosynthesis genes in deoxynivalenol- and nivalenol-producing *Fusarium graminearum* isolates. Mycological Research. 2003;**107**:190-197

[37] Brown DW, McCormick SP, Alexander NJ, Proctor RH, Desjardins AE. A genetic and biochemical approach to study trichothecene diversity in *Fusarium sporotrichioides* and *Fusarium graminearum*. Fungal Genetics and Biology. 2001;**32**(2):121-133

[38] Chandler EA, Simpson DR, Thomsett MA, Nicholson P. Development of PCR assays to *Tri7* and *Tri13* trichothecene biosynthetic genes, and characterisation of chemotypes of *Fusarium graminearum*, *Fusarium culmorum* and *Fusarium cerealis*. Physiological and Molecular Plant Pathology. 2003;**62**:355-367

[39] Waalwijk C, Kastelein P, De Vries I, Kerenyi Z, Van der Lee T, Hesselink T. Major changes in *Fusarium* spp. in wheat in the Netherlands. European Journal of Plant Pathology. 2003;**109**:743-754

[40] Jennings P, Coates ME, Turner JA, Chandler EA, Nicholson P. Determination of deoxynivalenol and nivalenol chemotypes of *Fusarium culmorum* isolates from England and Wales by PCR assay. Plant Pathology. 2004;**53**:182-190

[41] Jennings P, Coates ME, Walsh K, Turner JA, Nicholson P. Determination of deoxynivalenol- and nivalenol-producing chemotypes of *Fusarium graminearum* isolated from wheat crops in England and Wales. Plant Pathology. 2004;**53**:643-652

[42] Li HP, Wu AB, Zhao CS, Scholten O, Löffler H, Liao YC. Development of a generic PCR detection of deoxynivalenol- and

nivalenol-chemotypes of *Fusarium graminearum*. FEMS Microbiology Letters. 2005;**243**:505-511

[43] Quarta A, Mita G, Haidukowski M, Santino A, Mulè G, Visconti A. Assessment of trichothecene chemotypes of *Fusarium culmorum* occurring in Europe. Food Additives and Contaminants. 2005;**22**:309-315

[44] Bakan B, Giraud-Delville C, Pinson L, Richard-Molard D, Fournier E, Brygoo Y. Identification by PCR of *Fusarium culmorum* strains producing large and small amounts of deoxynivalenol. Applied and Environmental Microbiology. 2002;**68**:5472-5479

[45] Quarta A, Mita G, Haidukowski M, Logrieco A, Mulè G, Visconti A. Multiplex PCR assay for the identification of nivalenol, 3- and 15-acetyl-deoxynivalenol chemotypes in *Fusarium*. FEMS Microbiology Letters. 2006;**259**:7-13

[46] Wang JH, Li HP, Qu B, Zhang JB, Huang T, Chen FF, et al. Development of a generic PCR detection of 3-acetyldeoxy-nivalenol-, 15-acetyldeoxynivalenol- and nivalenol-chemotypes of *Fusarium graminearum* clade. International Journal of Molecular Sciences. 2008;**9**:2495-2504

[47] Suzuki F, Koba A, Nakajima T. Simultaneous identification of species and trichothecene chemotypes of *Fusarium asiaticum* and *F. graminearum sensu stricto* by multiplex PCR. Journal of General Plant Pathology. 2010;**76**:31-36

[48] Zhang H, Zhang Z, Van der Lee T, Chen WQ, Xu J, Xu JS, et al. Population genetic analyses of *Fusarium asiaticum* populations from barley suggest a recent shift favoring 3-ADON producers in Southern China. Phytopathology. 2010;**100**:328-336

[49] Wang JH, Zhang JB, Chen FF, Li HP, Ndoye M, Liao YC. Amultiplex PCR assay for genetic chemotyping of toxigenic *Fusarium graminearum* and wheat grains for 3-acetyldeoxynivalenol, 15-acetyldeoxynivalenol and nivalenol mycotoxin. Journal of Food, Agriculture and Environment. 2012;**10**:505-511

[50] Kulik T. Development of TaqMan assays for 3ADON, 15ADON and NIV *Fusarium* genotypes based on *Tri12* gene. Cereal Research Communications. 2011;**39**:200-214

[51] Nielsen LK, Jensen JD, Rodríguez A, Jørgensen LN, Justesen AF. *TRI12* based quantitative real-time PCR assays reveal the distribution of trichothecene genotypes of *F. graminearum* and *F. culmorum* isolates in Danish small grain cereals. International Journal of Food Microbiology. 2012;**157**:384-392

[52] Pasquali M, Giraud F, Brochot C, Cocco E, Hoffman L, Bohn T. Genetic *Fusarium* chemotyping as a useful tool for predicting nivalenol contamination in winter wheat. International Journal of Food Microbiology. 2010;**137**:246-253

[53] Berthiller F, Crews C, Dall'Asta C, Saeger SD, Haesaert G, Karlovsky P, et al. Masked mycotoxins: A review. Molecular Nutrition & Food Research. 2013;**57**:165-186

[54] Gratz SW. Do plant-bound masked mycotoxins contribute to toxicity? Toxins. 2017;**9**:85. DOI: 10.3390/toxins9030085

Chapter 6

Recent Biosensors Technologies for Detection of Mycotoxin in Food Products

Kobun Rovina, Sulaiman Nurul Shaeera, Joseph Merrylin Vonnie and Su Xin Yi

Abstract

Mycotoxins are chemically diverse and capable of inducing a wide diversity of acute and chronic symptoms, ranging from feed refusal to rapid death. Accurate detection and monitoring of mycotoxins is an essential component of the prevention, diagnosis, and remediation of mycotoxin-related issues in livestock and human food. Current trends in food analysis are focusing on the application of fast, simple procedure needed, and low-cost biosensor technologies that can detect with high sensitivity and selectivity different compounds associated with food safety. This chapter discussed the recent analytical methods-based biosensor technology for quantification of mycotoxins in food products. Mainly focus on the biosensor technology based on the immobilization of antibodies onto various nanomaterials such as nanoparticles, graphite, carbon nanotubes, and quantum dots. The nanomaterials are able to be functionalized with various biomolecules such as enzymes, antibodies, nucleic acids, DNA/RNA aptamers, bio- or artificial receptors that make them suitable for detection of various substances such as food toxins, bacteria, and other compounds important in food analysis. All the nanomaterials provide an effective platform for achieving high sensitivity that is similar and, in some cases, even better than conventional analytical methods. We believe that future trends will be emphasized on improving biosensor properties toward practical application in the food industry.

Keywords: mycotoxin, biosensor, nanomaterials, analytical methods, fungi

1. Introduction

Fungi are an organism that exists either in single-celled or complex multicellular organisms. This number of the organism may cause diseases by producing toxic substances which known as mycotoxins. Mycotoxins are toxic secondary metabolites of various fungi that significantly impact global food safety and security, from toxin exposure, economic loss of crops, or the salability of said crops. They are a widespread mixture of contaminants in various agricultural and food products, with both acute and chronic toxicological effects on human health [1]. Mycotoxin produced mainly by mycelial structure of filamentous fungi or specifically molds that may cause a harmful effect to animals as well as humans such as carcinogenic, nephrotoxicity, mutagenic, immunosuppressive, estrogenic neurotoxicity, reproductive and developmental toxicity, hepatotoxicity and indigestion [2].

Mycotoxins including aflatoxins (AFs), ochratoxins (OT), trichothecenes, zearalenone (ZEN), fumonisins (F), tremorgenic toxins, and ergot alkaloids mostly affect the public health and agro-economic significance. Factors affecting the magnitude of toxicity to the living organism are by consuming mycotoxin-contaminated foods or feeds, including species, mechanisms/modes of action, metabolism, and defense mechanisms [3]. Most of the countries agreed to set the limits of mycotoxins present in food because of the effects of the mycotoxins to human health. The permitted level is slightly different, which depends on the type of food products. The minimum limits for mycotoxins in single ppb (part per billion) and even below (0.05 ppb for infant foods) are established in EU, with similar standards in China and Japan [4].

Guan et al. [5] reported about 98% of the agricultural commodities, including corn, compound animal feeds, silage, cornmeal, puffed corn, wheat, bran, soybean meal, rapeseed meal, cottonseed meal and whole cottonseed content various group of mycotoxins. Besides, Smith et al. [6] stated that several mycotoxins contaminate approximately 48% of 7049 feedstuffs. Thus, it is essential to detect mycotoxins in the food industry to address the mycotoxin-related health issues to humans and animals effectively.

Conventional techniques such as thin-layer chromatography (TLC), high-performance liquid chromatography (HPLC) and mass spectrometry have been suggested by international organizations as standard approaches to study the occurrence of mycotoxins in food products [7]. Besides, enzyme-linked immunosorbent assay (ELISA) had been widely used to identify different types of mycotoxins. However, it has slight defects of cross-reactivity and possible false-positive or false-negative outcomes [8]. Also, those techniques usually costly and available in a specialized research laboratory needs highly personnel trained and laborious. Recently, advanced methods used to detect the presence of mycotoxins in food samples, which show high sensitivity, low cost, simple operation, and portable on-field use [9]. Besides, portable and easy-to-use biosensor devices suitable for express, in-field detection of mycotoxins. The development of biosensors for mycotoxins has risen sharply in the last decade with a large number of different bio-sensing technologies application. Zheng et al. (2006) reported biosensor as rapid methods which typically cost-effective, easy to be handled as well as a portable device to be used in an interchanging site compared to laboratory analysis.

2. Mycotoxin

Fungal toxins are secondary metabolites, which can cause some diseases in living things known as mycoses; meanwhile, dietary exposure to such metabolites produces the disease named mycotoxicoses. Mycotoxins are known as secondary metabolites, produced from microfungi and able to cause–effect human health as well as animals. Mycotoxins are commonly used as antibiotics and growth promotants because of their unique characteristics in pharmacological activity. Most of the mycotoxin are found as natural contaminant food, mainly in vegetable and feed. Nut, cereals, oilseeds, dried fruits, spices, and food from animal origins for example milk, egg, and meat are also may contain mycotoxin either outside or inside the product [10, 11]. A mycotoxin is believed no function in the life of a producer cell, unlike primary metabolites [12]. There are few types of mycotoxin such as aflatoxins (AFs), zearalenone (ZEA), deoxnivalenol (DON), ochratoxin (OTA) and T-2 toxin (trichothecene mycotoxin) which are a significant threat to the life and health of human and live stocks [13]. Mycotoxins are low molecular weight and thermal-stable secondary metabolite of toxic molds that belong to genera *Aspergillus,*

Penicillium, Alternaria, and *Fusarium*. These toxins are present in the mycelium and spore of the mold. Mycotoxin may become a biological weapon in bioterrorism because of its acute and chronic toxicities [14].

3. Types of mycotoxin

The established mycotoxins for agriculture and public health concerns including aflatoxins, ochratoxins, zearalenone, T-2 and HT-2 toxin, deoxynivalenol, fumonisins, citrinin, patulin, and ergot alkaloids shown in **Figure 1**. Aflatoxins B1 and M1 (AFT B1 & M1) [15] produced by *Aspergillus flavus* and *A. parasiticus* species grown on grains and cereals, spices, tree nuts. Aflatoxin B1(AFB1) is one of the most carcinogenic substances produced by fungi and results in inevitable contamination of food and feed at deficient concentrations. Four main types of aflatoxin naturally contaminate foods which are aflatoxin B1 (AFB1), G1 (AFG1) and their dihydroderivatives B2 (AFB2) and G2 (AFG2). Others without additional metabolites known as Aflatoxin M1 and Aflatoxin M2 [16]. AFT M1 being a 4-hydroxylated metabolite of AFT B1, is found in cow and sheep milk and milk products. Some studied had been identified there is 20 aflatoxins that belongs to a group called highly substituted difuranocoumarins. The International Agency for Research on Cancer (IARC) had been classified aflatoxin as very toxic compounds in group 1 due to evidence that shows the carcinogenicity in human [17].

Ochratoxin A (OTA) produced by *Aspergillus ochraceus, A. carbonarius,* and *Penicillium verrucosum* is one of the most abundant contaminants in grain and pork products, coffee, dried grapes, as well in wine and beer at humidity around 15–19%

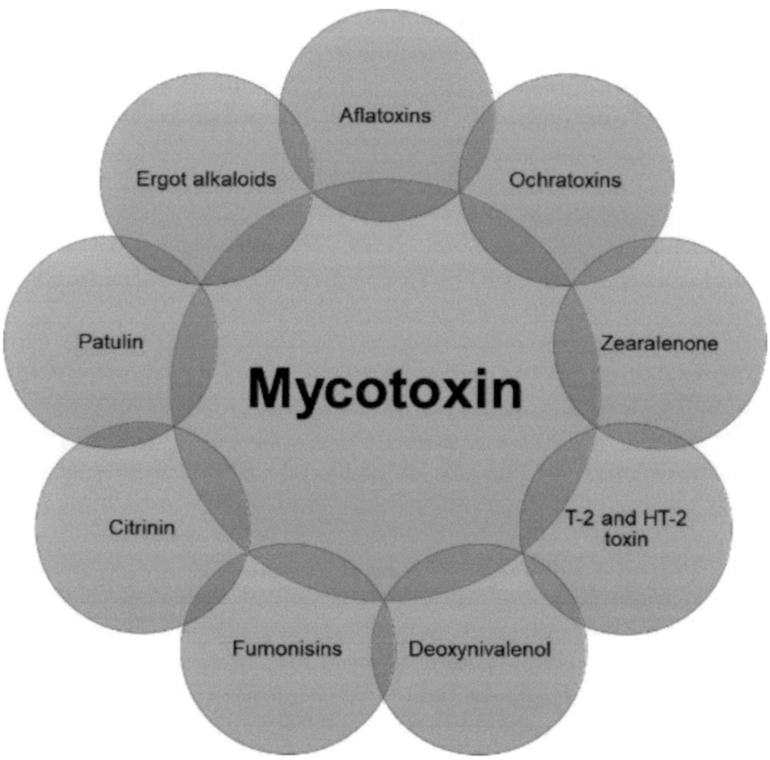

Figure 1.
Primary groups of mycotoxins in various food products.

and temperature ≥15°C [18]. OTA is carcinogenic and neurotoxic for humans, and immunotoxic for animals [19]. OTA can cause various forms of kidney, liver, and brain diseases in both humans and animals, although the trace amount of OTA usually is present in food [20].

Zearalenone (ZEN) produced by *Fusarium* or *Giberella* species grown on crops (maize, barley, oats, wheat, rice, also bread) is a potent estrogen metabolite causing infertility in swine and poultry [21].

4. Isolation of a mycotoxin from real samples

4.1 Solid-phase extraction (SPE)

A variation of chromatographic techniques based on small disposable cartridges packed with silica gel or bonded phase, which in the stationary phase is the basic principle of solid-phase extraction. The sample loaded in one solvent under low pressure and rinsed to remove the most of contaminant are moved and eluted in another solvent. These cartridges have a high capacity for small binding molecules. Different bonding phase such as silica gel, aminopropyl, florisil, phenyl, ion exchange materials, anionic and cationic to affinity materials including immunoadsorbents and molecular imprinted polymers (MIPs) are available in SPE cartridges [22]. OTA formation occurs in some Spanish sweet, which going drying process. C-18 column had been shown successful recovery above 90% of OTA, which enables to be isolated from the matrix [23]. Silica gel frequently used in SPE because the surface of silica particles is heterogeneous with a variety of silanol group which can bind target compound through multiple electrostatic interactions. Generally, silica gel was used directly or after modification, and it is a hydrophobic phase which used in environmental and food analysis of toxin, which performed both polar and non-polar solvents. Previous research conducted by Leitner et al. [24] showed that the use of C-18 reverse-phase in the extraction of OTA from wine and offer good result with combination with mass spectroscopy.

4.2 Liquid: liquid extraction (LLE)

Liquid–liquid extraction (LLE) or also known as solvent extraction agitating different solubility of toxin in the aqueous phase and an immiscible organic phase to extract the compound into one solvent and leaving the rest of matrix in others phase. A solvent such as hexane and cyclohexane are used to remove non-polar contaminant or molecule, for example, lipids, and cholesterol [25]. The common goal of LLE is sample clean-up and analyte component pre-concentration. Sample clean-up requires high selectivity of partitioning analyte component over potential interferents while analyte component pre-concentration require high distribution ratio to analyte can be extracted from a large volume of sample too small volume of extractant. Two bulk-liquid phases at least which are an aqueous phase that contains dissolved sample an organic extractant phase. The variety of condition will decide either the agitated mixture become the dispersed phase and another continuous phase. The thermodynamic driving force is resulting from the movement of chemical species from one bulk phase to another in two ways either by the difference in chemical potential for neutral species or electrochemical for ionic species [26]. Lately, Ezekiel et al. [27] used acetonitrile/water/acetic acid 79:20:1, (v/v/v) in a 50 mL polypropylene for the metabolites extraction and determination of apparent recoveries.

4.3 Supercritical fluid extraction (SFE)

Supercritical fluid extraction (SFE) had been used for years for industrial-scale separation and isolation of variety compound. SFE also has been utilizing in the field of food science to isolate not only natural food component but also unnatural compound like organic contaminants. SFE was developing and used as an alternative to extraction using liquid solvents. SFE considered an up-and-coming technique for the future because supercritical fluids have useful physical properties such as high viscosity and high diffusion constant for sample extraction which result in faster mass transport than regular and shorter the time for extraction. Using compressible gas like carbon dioxide (CO_2), the solvation power can be changed by altering the density or decrease the pressure to atmospheric pressure [28].

Most common supercritical fluid (SF) used is SC-CO_2, which is a suitable substituent for halogenated solvents. This is because the carbon dioxide is non-toxic, non-flammable, not significantly contribute to global warming and might be the cheapest solvent except for water. The usage of SFE to extract mycotoxin are very limited until recently because of the relative polar nature of mycotoxin and relative non-polar nature of food commodities such as nut and nut product. Taylor et al. [29] investigated the use of analytical SFE to remove aflatoxin Bi from field inoculated corn samples. Modification using a combination of various pressures "(2000-15,000 psi), temperatures (40–80°C), the quantity of SC-CO_2 (50–500 ml), and organic modifiers were used to optimize the extraction method. Optimal conditions were 5000 psi at 80°C with 15% modifier (acetonitrile/methanol 2:1) and a liquid carbon dioxide volume of 100 ml. The result gained from the extraction was 94.6% (RSD 6.2%, n = 5) of aflatoxin Bi could be recovered from ground corn contaminated at a level of approximately 500 μg/kg when using these settings.

5. Advanced techniques for detection of mycotoxin based biosensor

The integration of bioreceptors, nanomaterials, and different read-out techniques is capable of accomplishing the rapid, sensitive, and multiplexed detection of mycotoxins. In this section, the advanced applications of different read-out biosensors, including optical, EC, mass-sensitivity, and surface-enhanced Raman spectroscopy biosensors, integrated with the bio-receptors above and nanomaterials, are discussed (**Figure 2**).

5.1 Electrochemical biosensors

A biosensor is an analytical device that incorporating a bio-component or bio-receptor such as isolated enzymes, whole cell, tissues, aptamers with a suitable transducing system to detect chemical compound [30]. Measurement of the signal is generally electrochemical for biological, and this bio-electrochemical serves as transduction component in electrochemical biosensors. The biological reaction generates change in signal for conductance or impedance, measurable current or change accumulation, which can be measured by conductometric, potentiometric, or amperometric techniques [31]. The interaction between the target molecule and the electrical signal of bio-component produced can be measured.

Electrochemistry has been widely used in various fields, due to their high selectivity and sensitivity, high signal-to-noise ratio, simplicity, miniaturization, low cost, robust to liquid samples and more feasible for on-site application [20]. The electrochemical technique requires a reference, auxiliary, and a working electrode. Two exciting compounds are analyzed using compound biosensors that have interest

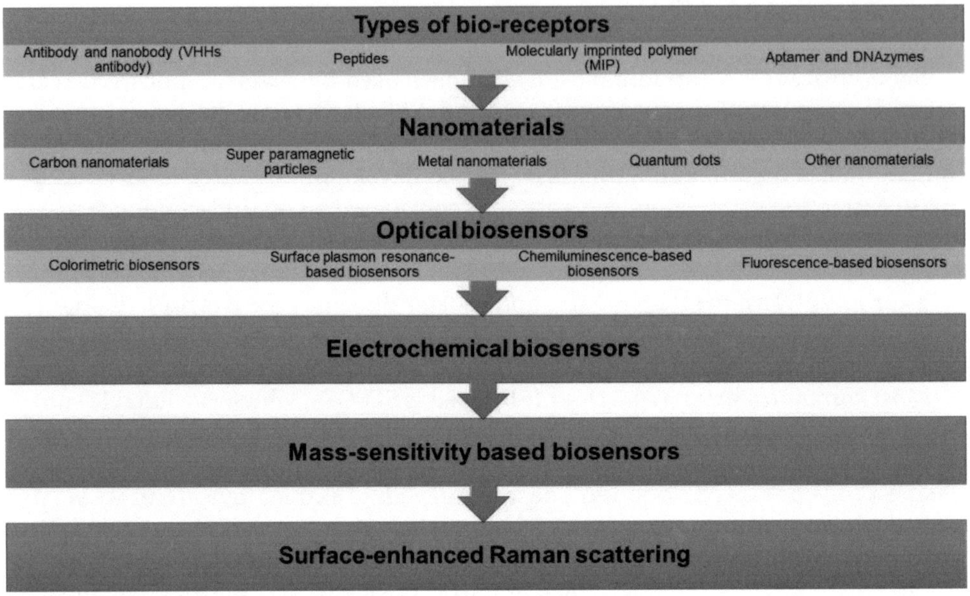

Figure 2.
The applications of different read-out biosensors integrated with bioreceptors and nanomaterials.

for nutritional food quality and contaminant such as toxin or pathogen that supposed not to be found in food products [30]. Selection of suitable working electrode is a crucial part of successful electrochemical measurement either by modification in working electrode materials or traditional metals such as mercury or gold [32].

Due to the widely occurring co-contamination of mycotoxins in raw food materials, Lu and Gunasekaran [33] designed and fabricated of an electrochemical immunosensor for simultaneous detection of two mycotoxins, fumonisin B1 (FB1) and deoxynivalenol (DON), in a single test. A dual-channel three-electrode electrochemical sensor pattern was etched on a transparent indium tin oxide (ITO)-coated glass via photolithography and was integrated with capillary-driven polydimethylsiloxane (PDMS) microfluidic channel. The achieved detection limits found 97 and 35 pg./mL, respectively. Besides, Nieto et al. [34] A third-generation enzymatic biosensor were developed to quantify sterigmatocystin (STEH). It was based on a glassy carbon electrode modified with a composite of the soybean peroxidase enzyme (SPE) and chemically reduced graphene oxide. A third-generation enzymatic biosensor to quantify STEH in corn samples spiked with the mycotoxin. The biosensor was based on glassy carbon (GC) electrode modified with a composite of SPE and chemically reduced graphene oxide (CRGO). The biosensor was also used to determine STEH in corn samples inoculated with Aspergillus flavus, which is an aflatoxins fungus producer. The biosensor showed a linear response in the concentration range from 6.9×10^{-9} to 5.0×10^{-7} mol L^{-1}. The limit of detection was 2.3×10^{-9} mol L^{-1} for a signal: noise ratio of 3:1.

5.2 Aptasensor

The aptamer is referred to the Latin word, aptus means "to fit," which relationship between aptamers and their target look like "lock-and-key" theory [35]. Aptamers usually single-stranded RNA or DNA, which consist of 2–60 nucleotides, which specifically bind to the target, including organic molecules and cells. Aptasensors referred to biosensors using aptamers as biorecognition element and aptasensor were described in 1996 [36] which had been used in multiple sensing applications.

Advantages using aptamers are aptamers can provide high stability and affinity. Aptamers also provide simplicity, low cost, and excellent batch-to-batch reproducibility. Aptasensor can attract massive attention because of excellent binding constant toward most mycotoxins. The critical step in the design of biosensors is immobilization of aptamers because this factor can affect the affinity of the aptamer for target and long-term stability for real sample. There are several immobilization strategies affect the used for aptasensor development. Firstly, the adsorption or π-π interaction between DNA bases aptamer and graphene oxide (GO)-modified interfaces [37]. The covalent linkage of the aptamer to the carboxylic acid group that presents on surface or nanomaterial [38] and thiolated binding aptamers to CdTe quantum dots (QDs) or Au-based materials [39]. Besides, affinity binding based on biotin-streptavidin or other affinity interaction [40, 41] and hybridization of partially complementary single-stranded DNA which immobilized on surface or nanoparticle [42]. Duan et al. [43] developed multicolor quantum dot nanobeads for simultaneous qualitative immunochromatographic detection of mycotoxins (ZEN, OTA, and FB_1) in corn samples with detection limits reached up to 5, 20, and 10 ng/mL within 10 min, respectively.

5.3 Immunosensor

Immunosensors are devices based on the detection of analyte-antibody interaction. Three main groups have been developing, which are luminescent or colorimetric sensors, surface plasmon resonance, and electrochemical sensors. The sensor usually combined with simple methanol–water for the extraction of a mycotoxin from food samples. Colorimetric and luminescent are based on the visible or UV light transformation into an analytical signal [44]. A colorimetric sensor developed for AFB1 detection using direct competitive ELISA principle. The color was detected and measured with spectrometer by reading absorbance at 620 nm. According to Garden and Strachen [45], this method could detect AFB1 as low as 0.2 ng/mL in artificially contaminated food material as compared to the sensitivity of a microtitre plate ELISA.

Surface plasmon resonance (SPR) is an optical phenomenon which used for measure changes on the surface of thin metal films (Au or Ag) under condition total internal reflection [46]. The sensitivity of SPR sensors and microtiter plate ELISAs were compared for detection of AFB1 using same immunoreagents, which are a polyclonal antibody and AFB1-BSA conjugate. As a result, the SPR sensor (3.0–49 ng/mL) is a more sensitive but narrow and linear range of detection compared to ELISA (12–25,000 ng/mL) [47]. Electrochemical immunosensor for mycotoxin are based on competitive ELISA principle, which electrochemical transducer allows detection redox directly [44]. Pemberton et al. [48] in their study, a calibration plot AFB1 obtained over the concentration range from 0.15 to 2.5 ng/mL, which give detection limit around 0.15 ng/mL in buffer solution.

OTA is small molecules that possess one epitope and no more than one antibody can bind due to their small molecular size. This molecule was detected using a competitive assay rather than a sandwich assay format. The competitive assay is based on the competition of immobilized antigen and a free antigen for the antibody in solution. One of the critical parameters to determine the sensitivity and limit of detection (LOD) is antibody concentration. The excessive antibody in solution may cause more antigen needed to create a measurable difference in signal. Therefore, to increase the binding capacity, protein conjugate such as SPR sensor development was used which the OTA either directly conjugated to BSA or PEG. The sensitivity increased with decreasing antibody concentration because the PEG-linked surface needs less initial antibody concentration for efficient analysis. Pirincci et al. [49]

described that the OTA-sensitive QCM sensor was developed by direct immobilization of OTA to the sensor surface.

5.4 Molecularly imprinted polymer (MIP)

Molecular imprinted polymer (MIP) is a method which is described as a method that highly efficient in producing functional material that able to mimic natural recognition entities, such as antibodies and biological receptors [50] which equipped with particular identification characteristics. In 2009, an electrochemical sensor was built by Pardieu et al. [51] for the method of detection. Thus, this method is used to recognize a specific element for template molecule detection.

MIP is used in various field of application to recognize biological and chemical molecules including amino acids and proteins [52], nucleotide derivatives, pollutants, drugs and foods [53]. Molecularly imprinted polymer method had been applied in chromatography for HPLC and GC, Solid phase extraction, Chemical sensor systems, catalysis, drug delivery, antibodies, and receptors system [54]. The formation of a complex between an analyte and the functional monomer determines the Molecularly imprinted polymer. A three-dimensional polymer network is formed due to the presence of a significant excess of a cross-linking agent [55]. A specific recognition site is formed which complementary in shape, size, and chemical functionality to the template molecule as the template being removed from the polymer after the polymerization process occurs as shown in the figure. The recognition phenomena occur when the intermolecular interactions such as hydrogen bonds, dipole–dipole, and ionic interactions between the template molecule and the functional groups present in the polymer matrix. This method is used due to their high selectivity and affinity for the target molecules. Therefore, the recognized polymer will bind to the template molecule only selectively.

The molecularly imprinted materials have excellent physical and chemical characteristics. The materials can resist high physical and chemical reaction against external degrading factor. Thus, the molecularly imprinted polymer is stable against mechanical stress, high temperature, and pressure, resistant against treatment with acid, base, or metal ions, and also stable in a wide range of solvents [56]. Sellergren firstly reported the application of MIP in solid phase extraction in 1994. Generally, the MIP as a sorbent was recognized as an accurate, selective, and sensitive pre-treatment method in detecting trace amounts of chemicals in the matrix. The application of MIP in solid phase extraction is used for veterinary residues, pesticides residue, illegal drugs, mycotoxins, and persistent organic pollutants had been published.

5.5 Optical biosensors

Biosensors can be divided into different groups, which are electrochemical, optical, thermometric, piezoelectric, or magnetic [57]. Somehow, the optical biosensor is the most preferred among the other methods. This is because it has powerful analytical techniques which have a high specification, sensitivity, small size, and cost-effectiveness [58, 59]. An optical biosensor is a device which is selective and sensitive that can detect deficient levels of chemicals and biological substances and for the measurement of molecular interactions in situ and in real time [60].

Optical methods, such as colorimetric, fluorescent, chemiluminescent, and surface plasmon resonant strategies, are proper techniques for mycotoxins detection due to their simplicity, rapidity, reliability, and high sensitivity. An optical biosensor is a system which combined various entities in a single system such as sampling, a biosensor, a system for replenishing information, and a data analysis system which to implement a biological model that provides information for human or machine [57]. The biosensor systems are developed by crucial attributes, which are the integration of fluidics, electronics, separation technology, and biological sub-systems. An optical biosensor is a compact analytical device, having a biological sensing element, integrated or connected to an optical transducer system. In this method, the analyte of interest that binds to the complementary optical bio-recognition element is recognized as immobilized on a suitable optical substrate [61]. An electronic signal is produced which the magnitude of the frequency is proportional that correspond to the concentration of an analyte or a group of analytes, to which the element will bind is the objective of optical biosensors [62]. Meanwhile, enzyme, substrate, antibody, and nucleic acids are used as the primary biological materials in optical biosensor technology [57]. The detection usually relies on an enzyme system which converts the analytes to products catalytically and can be oxidized or reduced at a working electrode.

Optical biosensing has two general modes, which are label-free and label-based. For label-free mode, the interaction of the analyzed material with the transducer will generate a detectable signal. On the contrary, the use of the label and the optical signal then generated by a colorimetric, fluorescent, or luminescent method are involved in label-based sensing [63]. The usage of optical biosensor depends on the different fields of use. This is because it has own requirements in term of measuring analysis, required precision of output, the sample concentration required, the time taken to complete the probe, the time necessary to prepare and reuse the biosensor, and the cleaning requirements of the system [57].

In the food industry, this method is used for the direct detection of bacteria in products. Optical biosensor used to detect the changes of refractive indices as the cell bind to the receptor, which is immobilized on the transducer [49]. The advantages of using optical biosensors are their speed, immunity of signal to electrical or magnetic interference. Besides, it is highly sensitive, reproducible, and simple-to-operate analytical tools. Somehow, some instrumentation involved in this method high in cost. Nabok et al. [4] reviewed the recent progress in the development of novel optical biosensing technologies for the detection of mycotoxins indirect assay with either specific antibodies or aptamers.

5.6 Enzymatic inhibition

There are a variety of enzymes such as cholinesterase, urease, glucose oxidase and more that have been applied in an enzymatic inhibition analysis and this method is pretty standard [64]. According to Puiu et al. [65], Acetylcholinesterase (AChE) is the most commonly used enzyme, and the reason is it is susceptible toward mycotoxin which is becoming the preferred method for mycotoxin detection. This statement is also supported by [66], which stated that biosensors for Aflatoxin B1 (a type of mycotoxin) or AFB1, in short, is developed by using AChE due to the inhibitory effect of AFB1 to AChE enzymatic activity. Also, the inhibitory effect of mycotoxin is a reversible process due to the non-covalently binding nature to the enzyme [67]. Soldatkin et al. [68] stated that aflatoxin showed the highest sensitivity toward enzymatic inhibition method among the other groups of toxins. A past study conducted by Egbunike and Ikegwuonu [69] also suggests that usage of cholinesterase in biosensor method as the biological component is usable as AFB1 detector as aflatoxicosis has been reported to be correlated with a significant reduction of acetylcholine turnover in rat brain.

Based on the previous research, it is proven that AChE is inhibited by the AFB1 from binding at the external site, which is located at the active site gorge entrance located at the tryptophan residue. The inhibitory effect of the AFB1 can be seen by its action where the toxin blocks the entrance to the active site so that the substrate cannot enter to participate to the catalytic site result in the choline unable to exit as proposed by the steric blockade model [70]. Based on the observation in the study conducted by Hansmann et al. [71], their results lead them to two findings. The first observation is the addition of AFB1 in the binding site of the active site did not fulfill the description for inhibitory activity, and this suggests that the AFB1 does not slide to the catalytic site. As for the second observation, mutation of Trp321 to alanine in Dm-AChE put a stop on the inhibitory activity at 10 μM concentration, and AFB1 at a concentration of 100 μM does not inhibit Hu-BuChE enzymatic activity. Also, the researchers assumed that AFB1 could not enter into the active site due to its relatively big size, especially when considering the hydrophilic shell might be further increased in size. Due to this condition, aflatoxin is grouped as a ligand which binds on the external site of the cholinesterase [72].

5.7 Mimotope

Mimotope or also known as peptide-displaying phage or synthetic peptides [73] is now one of the most reliable methods that are used to identify epitopes which are detected by monoclonal antibodies which are antibodies that made by the same immune cell is given that they are clones of one single parent cell. Next, the usage of mimotope in mycotoxin detection involves the usage of peptides which are identified to be structurally not identical to the original epitope of mycotoxin but at least have the properties to mimic the epitope by binding to the antibodies [74]. Generally, this method shared instead of the same concept with enzymatic inhibition, which in this case, the mimotope will be the one that elicits antibody. Also, this method is beneficial when the original epitopes (example from a mycotoxin) are hard to be isolated and at the same time only available in minimal amount [75]. The first assay that using mimotope for detection is being done by Yuan et al. [76], where a mimotope is used to identify the mycotoxin deoxynivalenol.

A study has been conducted by Sellrie et al. [74] which aims to describe a competitive immunoassay for identification of hapten fluorescein by utilizing a monoclonal anti-fluorescein antibody B13-DE1 and a mimotope peptide which act by binding to the antibody. Based on their findings, the peptide mimotope was conjugated to horseradish peroxidase (HRP) which is then competing for binding to monoclonial antibody B13-DE1 with fluorescein. Based on the result, they have proven that mimotopes can be used to utilization in simple yet sensitive immune assays in order to quantitatively identify and determine substance with low molecular weights. As for the reliability and reproducibility, the assay was proved by validation data and found to be in the range which is described in the literature for conventional competitive immunoassays by Wild [77].

6. Advanced techniques for detection of mycotoxin based biosensor

During the last few decades, consumers have become more aware of health and food quality, consequently, research on food safety augmented. The variety of contaminants in many food products requires the development of high-throughput, real-time, and portable detection methods. The evaluation of the different mycotoxins residues in foodstuffs became an essential factor in guaranteeing the products' quality. Hence, it is essential to improve the analytical standards to detect and quantify the presence of a mycotoxin. The operation procedure should be simplified continuously for the convenience of users. The biosensor based nanotechnology can be extensively used in food contaminants monitoring and eventually become effectively routine analysis tools that could meet numerous challenges.

Acknowledgements

The authors would like to thank the Universiti Malaysia Sabah for the support of this study.

Conflict of interest

The authors declare no conflict of interest.

Author details

Kobun Rovina*, Sulaiman Nurul Shaeera, Joseph Merrylin Vonnie and Su Xin Yi
Faculty of Food Science and Nutrition, Universiti Malaysia Sabah, Kota Kinabalu, Sabah, Malaysia

*Address all correspondence to: rovinaruby@ums.edu.my

References

[1] IARC (International Agency for Research on Cancer). Mycotoxin control in low-and middle-income countries. IARC Working Group Report No. 9. 2015

[2] Pleadin J, Frece J, Markov K. Mycotoxins in food and feed. Advances in Food and Nutrition Research. Available online 6 March 2019. In Press, Corrected Proof. 2019

[3] Hussein HS, Brasel JM. Toxicity, metabolism, and impact of mycotoxins on humans and animals. Toxicology. 2001;**167**(2):101-134. DOI: 10.1016/S0300-483X(01)00471-1

[4] Nabok A, Al-Rubaye AG, Al-Jawdah AM, Tsargorodska A, Marty JL, Catanante G, et al. Novel optical biosensing technologies for detection of mycotoxins. Optics and Laser Technology. 2019;**109**:212-221. DOI: 10.1016/j.optlastec.2018.07.076

[5] Guan S, Gong M, Yin Y, Huang R, Ruan Z, Zhou T, et al. Occurrence of mycotoxins in feeds and feed ingredients in China. Journal of Food, Agriculture and Environment. 2011;**9**(2):163-167. DOI: 10.1186/2049-1891-5-37

[6] Smith MC, Madec S, Coton E, Hymery N. Natural co-occurrence of mycotoxins in foods and feeds and their in vitro combined toxicological effects. Toxins. 2016;**8**(4):94

[7] Ferreira I, Fernandes JO, Cunha SC. Optimization and validation of a method based in a QuEChERS procedure and gas chromatography–mass spectrometry for the determination of multi-mycotoxins in popcorn. Food Control. 2012;**27**(1):188-193. DOI: 10.1016/j.foodcont.2012.03.014

[8] Ran R, Wang C, Han Z, Wu A, Zhang D, Shi J. Determination of deoxynivalenol (DON) and its derivatives: Current status of analytical methods. Food Control. 2013;**34**(1):138-148. DOI: 10.1016/j.foodcont.2013.04.026

[9] Guan J, Wang YC, Gunasekaran S. Using L-arginine-functionalized gold nanorods for visible detection of mercury (II) ions. Journal of Food Science. 2015;**80**(4):N828-N833. DOI: 10.1111/1750-3841.12811

[10] Adeyeye SA. Fungal mycotoxins in foods: A review. Cogent Food and Agriculture. 2016;**2**(1):1213127. DOI: 10.1080/23311932.2016.1213127

[11] Freire FDCO, da Rocha ME. Impact of mycotoxins on human health. Fungal Metabolites. 2017;**1**:239-261. DOI: 10.1007/978-3-319-25001-4_21

[12] Goyal S, Ramawat KG, Mérillon JM. Different shades of fungal metabolites: An overview. Fungal Metabolites. 2016;**1**:1-29. DOI: 10.1007/978-3-319-25001-4_34

[13] Anfossi L, Baggiani C, Giovannoli C, D'Arco G, Giraudi G. Lateral-flow immunoassays for mycotoxins and phycotoxins: A review. Analytical and Bioanalytical Chemistry. 2013;**405**(2-3):467-480. DOI: 10.1007/s00216-012-6033-4

[14] Overy DP, Seifert KA, Savard ME, Frisvad JC. Spoilage fungi and their mycotoxins in commercially marketed chestnuts. International Journal of Food Microbiology. 2003;**88**(1):69-77. DOI: 10.1016/S0168-1605(03)00086-2Get

[15] Yin YN, Yan LY, Jiang JH, Ma ZH. Biological control of aflatoxin contamination of crops. Journal of Zhejiang University. Science. B. 2008;**9**(10):787-792. DOI: 10.1631/jzus.B0860003

[16] Yagati A, Chavan S, Baek C, Lee MH, Min J. Label-free impedance sensing of aflatoxin B1 with polyaniline nanofibers/Au nanoparticle electrode array. Sensors. 2018;**18**(5):1320. DOI: 10.3390/s18051320

[17] IARC Working Group on the Evaluation of Carcinogenic Risks to Humans. Chemical Agents and Related Occupations. IARC Monographs on the Evaluation of Carcinogenic Risks to Humans, 100(PT F), 9. 2012

[18] Bayman P, Baker JL. Ochratoxins: A global perspective. Mycopathologia. 2006;**162**(3):215-223. DOI: 10.1007/s11046-006-0055-4

[19] Mateo R, Medina Á, Mateo EM, Mateo F, Jiménez M. An overview of ochratoxin A in beer and wine. International Journal of Food Microbiology. 2007;**119**(1-2):79-83. DOI: 10.1016/j.ijfoodmicro.2007.07.029

[20] Alhamoud Y, Yang D, Kenston SS, Liu G, Liu L, Zhou H, et al. Advances in biosensors for the detection of ochratoxin A: Bio-receptors, nanomaterials, and their applications. Biosensors and Bioelectronics. 2019:111418. DOI: 10.1016/j.bios.2019.111418

[21] Desjardins AE, Proctor RH. Molecular biology of Fusarium mycotoxins. International Journal of Food Microbiology. 2007;**119**(1-2):47-50. DOI: 10.1016/j.ijfoodmicro.2007.07.024

[22] Katerere DR, Stockenström S, Shephard GS. HPLC-DAD method for the determination of patulin in dried apple rings. Food Control. 2008;**19**(4):389-392. DOI: 10.1016/j.foodcont.2007.04.015

[23] Hernández MJ, García-Moreno MV, Durán E, Guillén D, Barroso CG. Validation of two analytical methods for the determination of ochratoxin A by reversed-phased high-performance liquid chromatography coupled to fluorescence detection in musts and sweet wines from Andalusia. Analytica Chimica Acta. 2006;**566**(1):117-121. DOI: 10.1016/j.aca.2006.02.002

[24] Leitner A, Zöllner P, Paolillo A, Stroka J, Papadopoulou-Bouraoui A, Jaborek S, et al. Comparison of methods for the determination of ochratoxin A in wine. Analytica Chimica Acta. 2002;**453**(1):33-41. DOI: 10.1016/S0003-2670(01)01483-0

[25] Turner NW, Subrahmanyam S, Piletsky SA. Analytical methods for determination of mycotoxins: A review. Analytica Chimica Acta. 2009;**632**(2):168-180. DOI: 10.1016/j.aca.2008.11.010

[26] Tang S, Zhang H, Lee HK. Advances in sample extraction. Analytical chemistry. 2015;**88**(1):228-249

[27] Ezekiel CN, Sulyok M, Ogara IM, Abia WA, Warth B, Šarkanj B, et al. Mycotoxins in uncooked and plate-ready household food from rural northern Nigeria. Food and Chemical Toxicology. 2019;**128**:171-179. DOI: 10.1016/j.fct.2019.04.002

[28] Anklam E, Berg H, Mathiasson L, Sharman M, Ulberth F. Supercritical fluid extraction (SFE) in food analysis: A review. Food Additives and Contaminants. 1998;**15**(6):729-750. DOI: 10.1080/02652039809374703

[29] Taylor SL, King JW, Richard JL, Greer JI. Analytical-scale supercritical fluid extraction of aflatoxin B1 from field-inoculated corn. Journal of Agricultural and Food Chemistry. 1993;**41**(6):910-913. DOI: 10.1021/jf00030a014

[30] Rotariu L, Lagarde F, Jaffrezic-Renault N, Bala C.

Electrochemical biosensors for fast detection of food contaminants–trends and perspective. TrAC, Trends in Analytical Chemistry. 2016;**79**:80-87. DOI: 10.1016/j.trac.2015.12.017

[31] Bard AJ, Faulkner LR, Leddy J, Zoski CG. Electrochemical Methods: Fundamentals and Applications. New York: wiley; 1980

[32] Mishra G, Barfidokht A, Tehrani F, Mishra R. Food safety analysis using electrochemical biosensors. Food. 2018;**7**(9):141. DOI: 10.3390/foods 7090141

[33] Lu L, Gunasekaran S. Dual-channel ITO-microfluidic electrochemical immunosensor for simultaneous detection of two mycotoxins. Talanta. 2019;**194**:709-716. DOI: 10.1016/j.talanta.2018.10.091

[34] Nieto CD, Granero AM, Garcia D, Nesci A, Barros G, Zon MA, et al. Development of a third-generation biosensor to determine sterigmatocystin mycotoxin: An early warning system to detect aflatoxin B1. Talanta. 2019;**194**:253-258. DOI: 10.1016/j.talanta.2018.10.032

[35] Peltomaa R, Benito-Peña E, Moreno-Bondi MC. Bioinspired recognition elements for mycotoxin sensors. Analytical and Bioanalytical Chemistry. 2018;**410**(3):747-771. DOI: 10.1007/s00216-017-0701-3

[36] Pfeiffer F, Mayer G. Selection and biosensor application of aptamers for small molecules. Frontiers in Chemistry. 2016;**4**:25. DOI: 10.3389/fchem.2016.00025

[37] McKeague M, Bradley CR, Girolamo AD, Visconti A, Miller JD, DeRosa MC. Screening and initial binding assessment of fumonisin B1 aptamers. International Journal of Molecular Sciences. 2010;**11**(12):4864-4881. DOI: 10.3390/ijms11124864

[38] Sabet FS, Hosseini M, Khabbaz H, Dadmehr M, Ganjali MR. FRET-based aptamer biosensor for selective and sensitive detection of aflatoxin B1 in peanut and rice. Food Chemistry. 2017;**220**:527-532. DOI: 10.1016/j.foodchem.2016.10.004

[39] Chen X, Bai X, Li H, Zhang B. Aptamer-based microcantilever array biosensor for detection of fumonisin B-1. RSC Advances. 2015;**5**(45):35448-36452. DOI: 10.1039/C5RA04278J

[40] Guo X, Wen F, Zheng N, Li S, Fauconnier ML, Wang J. A qPCR aptasensor for sensitive detection of aflatoxin M 1. Analytical and Bioanalytical Chemistry. 2016;**408**(20):5577-5584. DOI: 10.1007/s00216-016-9656-z

[41] Samokhvalov AV, Safenkova IV, Eremin SA, Zherdev AV, Dzantiev BB. Use of anchor protein modules in fluorescence polarisation aptamer assay for ochratoxin A determination. Analytica Chimica Acta. 2017;**962**:80-87. DOI: 10.1016/j.aca.2017.01.024

[42] Wang S, Zhang Y, Pang G, Zhang Y, Guo S. Tuning the aggregation/disaggregation behavior of graphene quantum dots by structure-switching aptamer for high-sensitivity fluorescent ochratoxin A sensor. Analytical Chemistry. 2017;**89**(3):1704-1709. DOI: 10.1021/acs.analchem.6b03913

[43] Duan H, Li Y, Shao Y, Huang X, Xiong Y. Multicolor quantum dot nanobeads for simultaneous multiplex immunochromatographic detection of mycotoxins in maize. Sensors and Actuators B: Chemical. 2019;**291**:411-417. DOI: 10.1016/j.snb.2019.04.101

[44] GoryachevaIY, SaegerSD, EreminSA, Peteghem CV. Immunochemical methods for rapid mycotoxin detection: Evolution from single to multiple analyte screening: A review. Food Additives and Contaminants. 2007;**24**(10):1169-1183. DOI: 10.1080/02652030701557179

[45] Garden SR, Strachan NJ. Novel colorimetric immunoassay for the detection of aflatoxin B1. Analytica Chimica Acta. 2001;**444**(2):187-191. DOI: 10.1016/S0003-2670(01)01231-4

[46] Homola J, Yee SS, Gauglitz G. Surface plasmon resonance sensors. Sensors and Actuators B: Chemical. 1999;**54**(1-2):3-15. DOI: 10.1016/S0925-4005(98)00321-9

[47] Daly SJ, Keating GJ, Dillon PP, Manning BM, O'Kennedy R, Lee HA, et al. Development of surface plasmon resonance-based immunoassay for aflatoxin B1. Journal of Agricultural and Food Chemistry. 2000;**48**(11):5097-5104. DOI: 10.1021/jf9911693

[48] Pemberton RM, Pittson R, Biddle N, Drago GA, Hart JP. Studies towards the development of a screen-printed carbon electrochemical immunosensor array for mycotoxins: A sensor for aflatoxin B1. Analytical Letters. 2006;**39**(8):1573-1586. DOI: 10.1080/00032710600713289

[49] Pirincci SS, Ertekin O, Laguna DE, Ozen FS, Guloglu FB, Ozturk ZZ, et al. Label-free QCM Immunosensor for the detection of ochratoxin A. Multidisciplinary Digital Publishing Institute Proceedings. 2017;**1**(8):706. DOI: 10.3390/s18041161

[50] Poma A, Turner AP, Piletsky SA. Advances in the manufacture of MIP nanoparticles. Trends in Biotechnology. 2010;**28**(12):629-637. DOI: 10.1016/j.tibtech.2010.08.006

[51] Pardieu E, Cheap H, Vedrine C, Lazerges M, Lattach Y, Garnier F, et al. Molecularly imprinted conducting polymer based electrochemical sensor for detection of atrazine. Analytica Chimica Acta. 2009;**649**(2):236-245. DOI: 10.1016/j.aca.2009.07.029

[52] Bossi A, Bonini F, Turner AP, Piletsky SA. Molecularly imprinted polymers for the recognition of proteins: The state of the art. Biosensors and Bioelectronics. 2007;**22**(6):1131-1137. DOI: 10.1016/j.bios.2006.06.023

[53] Baggiani C, Anfossi L, Giovannoli C. Solid phase extraction of food contaminants using molecular imprinted polymers. Analytica Chimica Acta. 2007;**591**(1):29-39. DOI: 10.1016/j.aca.2007.01.056

[54] Ge Y, Turner AP. Molecularly imprinted sorbent assays: Recent developments and applications. Chemistry—A European Journal. 2009;**15**(33):8100-8107. DOI: doi.org/10.1002/chem.200802401

[55] Ramström O, Mosbach K. Synthesis and catalysis by molecularly imprinted materials. Current Opinion in Chemical Biology. 1999;**3**(6):759-764. DOI: 10.1016/S1367-5931(99)00037-X

[56] Kriz D, Ramstroem O, Svensson A, Mosbach K. A biomimetic sensor based on a molecularly imprinted polymer as a recognition element combined with fiber-optic detection. Analytical Chemistry. 1995;**67**(13):2142-2144. DOI: 10.1021/ac00109a037

[57] Dey D, Goswami T. Optical biosensors: A revolution towards quantum nanoscale electronics device fabrication. BioMed Research International. 2011;**2011**:7. Article ID 348218. DOI: 10.1155/2011/348218

[58] Luo XL, Xu JJ, Zhao W, Chen HY. Glucose biosensor based on ENFET doped with SiO_2 nanoparticles. Sensors and Actuators B: Chemical. 2004;**97**(2-3):249-255. DOI: 10.1016/j.snb.2003.08.024

[59] Sant W, Pourciel ML, Launay J, Do Conto T, Martinez A, Temple-Boyer P. Development of chemical field effect transistors for the detection of urea. Sensors and Actuators B: Chemical. 2003;**95**(1-3):309-314. DOI: 10.1016/S0925-4005(03)00430-1

[60] Nath N, Chilkoti A. A colorimetric gold nanoparticle sensor to interrogate biomolecular interactions in real time on a surface. Analytical Chemistry. 2002;**74**(3):504-509. DOI: 10.1021/ac015657x

[61] Li M, Chen L, Zhang W, Chou SY. Pattern transfer fidelity of nanoimprint lithography on six-inch wafers. Nanotechnology. 2002;**14**(1):33. DOI: 10.1088/0957-4484/14/1/308

[62] Turner DC, Chang C, Fang K, Brandow SL, Murphy DB. Selective adhesion of functional microtubules to patterned silane surfaces. Biophysical Journal. 1995;**69**(6):2782-2789. DOI: 10.1016/S0006-3495(95)80151-7

[63] Damborský P, Švitel J, Katrlík J. Optical biosensors. Essays in biochemistry. 2016;**60**(1):91-100

[64] Chauhan R, Singh J, Sachdev T, Basu T, Malhotra BD. Recent advances in mycotoxins detection. Biosensors and Bioelectronics. 2016;**81**:532-545. DOI: 10.1016/j.bios.2016.03.004

[65] Puiu M, Istrate O, Rotariu L, Bala C. Kinetic approach of aflatoxin B1–acetylcholinesterase interaction: A tool for developing surface plasmon resonance biosensors. Analytical Biochemistry. 2012;**421**(2):587-594. DOI: 10.1016/j.ab.2011.10.035

[66] Moscone D, Arduini F, Amine A. A rapid enzymatic method for aflatoxin B detection. In: Microbial Toxins. Totowa, NJ: Humana Press; 2011. pp. 217-235. DOI: 10.1007/978-1-61779-102-4_20

[67] Stepurska KV, Soldatkin OO, Kucherenko IS, Arkhypova VM, Dzyadevych SV, Soldatkin AP. Feasibility of application of conductometric biosensor based on acetylcholinesterase for the inhibitory analysis of toxic compounds of different nature. Analytica Chimica Acta. 2015;**854**: 161-168. DOI: 10.1016/j.aca.2014.11.027

[68] Soldatkin OO, Burdak OS, Sergeyeva TA, Arkhypova VM, Dzyadevych SV, Soldatkin AP. Acetylcholinesterase-based conductometric biosensor for determination of aflatoxin B1. Sensors and Actuators B: Chemical. 2013;**188**:999-1003. DOI: 10.1016/j.snb.2013.06.107

[69] Egbunike GN, Ikegwuonu FI. Effect of aflatoxicosis on acetylcholinesterase activity in the brain and adenohypophysis of the male rat. Neuroscience Letters. 1984;**52**(1-2):171-174. DOI: 10.1016/0304-3940(84)90369-0

[70] Szegletes T, Mallender WD, Rosenberry TL. Nonequilibrium analysis alters the mechanistic interpretation of inhibition of acetylcholinesterase by peripheral site ligands. Biochemistry. 1998;**37**(12):4206-4616. DOI: 10.1021/bi972158a

[71] Hansmann T, Sanson B, Stojan J, Weik M, Marty JL, Fournier D. Kinetic insight into the mechanism of cholinesterasterase inhibition by aflatoxin B1 to develop biosensors. Biosensors and Bioelectronics. 2009;**24**(7):2119-2124. DOI: 10.1016/j.bios.2008.11.006

[72] Bourne Y, Taylor P, Radić Z, Marchot P. Structural insights into ligand interactions at the acetylcholinesterase peripheral anionic site. The EMBO Journal. 2003;**22**(1):1-2. DOI: 10.1093/emboj/cdg005

[73] Kramer A, Schneider-Mergener J. Synthesis and screening of peptide libraries on continuous cellulose membrane supports. In: Combinatorial Peptide Library Protocols. Vol. 87. Humana Press; 1998. pp. 25-39. DOI: 10.1385/0-89603-392-9:25

[74] Sellrie F, Schenk JA, Behrsing O, Böttger V, Micheel B. A competitive immunoassay to detect a hapten using

an enzyme-labelled peptide mimotope as tracer. Journal of Immunological Methods. 2002;**261**(1-2):141-144. DOI: 10.1016/S0022-1759(01)00561-0

[75] Kieber-Emmons T, Monzavi-Karbassi B, Wang B, Luo P, Weiner DB. Cutting edge: DNA immunization with minigenes of carbohydrate mimotopes induce functional anti-carbohydrate antibody response. The Journal of Immunology. 2000;**165**(2):623-627. DOI: 10.4049/jimmunol.165.2.623

[76] Yuan Q, Pestka JJ, Hespenheide BM, Kuhn LA, Linz JE, Hart LP. Identification of mimotope peptides which bind to the mycotoxin deoxynivalenol-specific monoclonal antibody. Applied and Environmental Microbiology. 1999;**65**(8):3279-3286. Available from https://www.ncbi.nlm.nih.gov/pmc/articles/PMC91492/

[77] Wild D. The Immunoassay Handbook. New York: Stockton Press; 1994

Chapter 7

The Potential Application of Nanoparticles on Grains during Storage: Part 1 – An Overview of Inhibition against Fungi and Mycotoxin Biosynthesis

Daniel Nsengumuremyi, Parise Adadi, Gavers K. Oppong, Nadezhda V. Barakova and Elena F. Krivoshapkina

Abstract

Cereals and legumes are the major staples across the globe, thus providing nutrition to humans, and their by-products utilized as animal feeds. However, mycotoxins synthesized by fungi contaminate these grains on the field during cultivation and are transferred to the storage centers. These fungi infect and deteriorate stored grains, thereby tampering with food security. Moreover, the deterioration decreases nutrient content and alters the physicochemical properties of grains. The current conventional methods used to reduce grain contamination are becoming ineffective, coupled with the detrimental health effects it has on the consumer and to the environment. Herein, we present an overview of the use of nanoparticles (NPs) as an alternative and novel method of reducing mycotoxin biosynthesis due to their potent biocidal properties. Silver nanoparticles (AgNPs) are considered and have shown promising and effective fungicidal properties against important storage fungi, and pests hence could be utilized in the agriculture and food sector for a vast myriad of applications. These may help to either minimize/eradicate the exposure to the mycotoxins and its adverse health effects, hence contributing to the holistic growth and development of people.

Keywords: grains, mycotoxins, nanoparticles, biocidal activities, reactive oxygen species

1. Introduction

According to [1], microbial contamination of grains has resulted in a decrease in its nutritional quality, therefore, negatively affecting the productivity of humans (the workforce of a nation). Grains (cereals and legumes) are staple foods and widely consumed around the world due to their nutritional value and calories. Eating food prepared from contaminated grains could lead to malnutrition due to insufficient nutrients in the grains or food poisoning from mycotoxins.

The presence of these mycotoxins affects the safety, quality, and functional properties of grains. Moreover, the organoleptic properties of products made from these grains could also be altered because some fungi strains produce potent odors, which serve as an antibiotic against other microorganisms [2]. There have been several reports regarding microbial contamination of grains [3–5] and the mycotoxins produced by some of these organisms potentially pose a health risk to consumers.

The ability of fungi to penetrate grains, reside within the endosperm, and utilize nutrients makes conventional methods insufficient to deal with the menace [6]. Therefore, the fundamental problem remains unsolved. A convenient and practical approach where the nutritional quality, sensorial properties, color, and shelf life of the grains remain unchanged is warranted in curbing this menace. Therefore, we propose nanoparticles as the ultimate solution to the predicament mentioned above since they are known to exert potent biocidal activities against the vast myriad of microorganisms [7–17] involves in contaminating grains, hence could be utilized as antifungal agents during grain storage.

This chapter summarizes the microbial contamination of grains and the existing conventional methods employed to curb and or minimize this menace. Also, the potential application of silver nanoparticles as an alternative to the traditional techniques is discussed.

1.1 The economic importance of grains

Foodgrains could be cereals or legumes (pulses). The world leading cereal grains are wheat, barley, rice, maize, oats, rye, millet, and sorghum. Reports show that cereals are the dominant crops cultivated globally, with 2500 million tons harvested in 2011. The proportion of maize, rice, and wheat harvested is 883; 723; 704 million tons, respectively [18, 19]. Cereals are whole, hulled, cracked, rolled, or ground forms of products produced from various grains constituting staple foods for many localities globally. They contain a substantial amount of starch, a carbohydrate that provides dietary energy [20]. Also, cereals are utilized in feeding livestock. Huntington [21] reported a starch content of 72% for corn and sorghum, while 57, 58 and 77% for barley, oat and wheat. Thus could be utilized to feed ruminants due to their high energy values. The role of cereal grains in the world food supply cannot be undermined as it provides 75% of the calories and protein in the human diet [22]. In Russia, folks use cereals in brewing (beer, kvass), production of distillates, and food (i.e., sweets, cookies, porridge, among others).

The second most important family of crops are the legumes, used for their grains, and as forage [23]. Previous works [24–26] have reported that legume seeds contain protein, soluble and insoluble fiber, slowly digested starch, micro- and macronutrients, and vitamins, in addition to various bioactive phytochemicals such as flavonoids and other antioxidants which are beneficial to human health. Legumes complement proteins in cereals and contain 20–45% protein compared to 7–17% in cereals [27]. Grain legumes are also utilized in feeding livestock, either as a concentrated compound feed (in poultry production) or as whole-crop forage (in cattle, sheep, and pig production) [23]. The presence of antinutritional factors (ANFs) such as Kunitz trypsin inhibitor (KTI), Bowman-Birk inhibitor, and lectins in legumes limits their utilization by humans and in animal husbandry with exception to ruminants (i.e., cattle, sheep and goat), which can degrade ANFs due to the microbial fermentation in their stomach [28]. ANFs can decrease the nutritive value of legumes and cause health problems that may be fatal for both humans (if a substantial amount is consumed) and animals [29]. Nevertheless, various methods have been proposed to decrease the concentration of these ANFs [30–32]. Legumes are also utilized in feeding fish, thus limit the need for expensive fishmeal in the pisciculture industry [33–35]. Therefore, the safety and quality of grain legumes ought to be screened before utilization to avoid any further complications due to ANFs and mycotoxins.

1.2 Sources of fungi contamination of grains

Microorganisms plays vital role in balancing the ecosystem; they aid in the digestion of food in humans; are utilized in the production of food (i.e., starter culture in brewing, cheese production, among others), and serve as a good source of vital enzymes (exogenous enzymes). Nevertheless, these microorganisms could cause problems such as food poisoning (due to some mycotoxins they secrete), food spoilage, and grain contamination.

The entire production process (sowing, harvesting, postharvest drying, and storage) of grains are possible sources of fungi contamination [36]. Dust, water, diseased plants, insects, soil, fertilizers, animal excreta, and environmental pollutants are possible origins of fungi cross contamination. The farmer, the processor, and the distributor could be a source of microbial contamination as well as contaminated farm machinery and unclean storage facilities (silos, etc.). According to [37], microbial contamination from the skin, mouth, and nose of food handlers could be directly introduce into the food chain. During drying, most farmers step on the grains with their Wellington boots, which is a possible route of introducing microorganism [38].

The microflora of grains mainly belong to the *Alternaria, Fusarium, Helminthosporium*, and *Cladosporium* families. Yeasts were isolated from grains; however, its load was less compared to mold [4]. Mechanical damage during harvesting or processing could serve as a route via which fungi could penetrate the endosperm of seeds, reproduce, and secrete mycotoxins (aflatoxins, etc.), rendering the food unsafe for human consumption. According to the International Commission on Microbiological Specifications for Foods [39], isolated fungi were mainly on the surface of the kernel; only a few species occupy the inner parts of the seeds due to damage. Birds could introduce fungi on grains by (1) feeding on crops in the field. This can introduce gut microbiota to these plants, which could subsequently be spread by rainwater. (2) Their feet could also aid to spread microbes by landing and picking up fungi spores from a diseased plant/crop to healthy ones. Bats, and insects (bees) could also aid the contamination of crops on the field, which can spread during harvesting.

According to [40], the primary cause of spoilage in stored grains in developed countries is attributed to fungi, because insects and rodents are controlled successfully. Factors such as high temperature, humidity, and poor storage conditions create a conducive environment for fungi to flourish and synthesize mycotoxins. These secondary metabolites can cause diseases in humans and animals. For instance, aflatoxins, ochratoxin A, deoxynivalenol, zearalenone, fumonisins, HT-2, and T-2 are classes of mycotoxins produced by various fungus species [41, 42]. Grapes were found to be contaminated with ochratoxin A, thus contaminating any product processed from them (juice, wine, vinegar, and dried grapes) [3].

2. Factors promoting microbial growth and mycotoxin production

When deciding whether moisture, temperature, etc., affects the safety of grains, other factors should be considered to settle on a scientifically proven conclusion. Extrinsic factors (temperature, relative humidity, mechanical injury on seeds during harvest or processing, insects, and rodents infestation) are environmental and physical factors surrounding the grains whereas those attributed to the characteristics of the grains are intrinsic factors (pH, acidity, nutrient composition, biological structure, moisture content/water activity, redox potential, naturally occurring and added antimicrobial factors). Details on how these factors contribute to or promote microbial contamination of grains are examined below.

2.1 Nutrient content

Every organism requires essential nutrients for growth and maintenance of metabolic functions. Hence, the type and concentration of nutrients needed depends on the class of microorganism. A source of energy, water, nitrogen, vitamins, minerals, and other compounds provide these nutrients. The growth of *Aspergillus flavus* on grains was significantly affected by the concentration of soluble sugars. Low sugar levels retarded its growth, whereas concentrations between 3.0 and 6.0% resulted in rapid growth, and the subsequent production of aflatoxin B1. Nevertheless, aflatoxin B1 production was significantly promoted due to the bioavailability of amino acids (arginine, glutamic acid, aspartic acid) and zinc in the grains [43]. In a similar study, Li et al. [44] reported different concentrations of mycotoxins (aflatoxin B1 (AFB), deoxynivalenol (DON), zearalenone (ZEA) and ochratoxin A (OTA)) on numerous swine feeds. These outcomes could be attributed to the nutritional composition of the feeds. The nutritional requirement of pigs depends on the state (gestating, finisher, grower, starter, etc.) hence varied feed rations are given which contain different nutrient concentration; as a result influence fungi growth and subsequent mycotoxins production. The bioavailability of nutrients in most grains would support the growth of a wide range of microorganisms. Although each strain of mold has the genetic potential to produce a particular mycotoxin, nutrient bioavailability could influence their levels significantly [45].

2.2 Biological structure

Grains have biological structures which prevent the penetration and growth of microorganisms. The testa of seeds and shell of nuts are examples of such structures. Some physical structures/barriers may exert antimicrobial potential. Intact biological structures prevent the entry of microbes, subsequent growth and production of mycotoxins in grains. However, these structures are destroyed during harvesting, transporting, or processing of the grains. Insect infestation could pave way for microbial proliferation of grains [46, 47]. Extract of Peanut testa was reported to exhibit pronounced antifungal activities against *Penicillium* sp., *A. niger*, and *Actinomucor* sp. The cardinal and purple peanut testa produced a significant zone of inhibition at concentrations of 0.8 and 2.0 g/L, respectively. It was concluded that the fungicidal potentials of the testa depend on the type of peanut [48]. Nevertheless, the environment, variety, type of farming system adopted, duration of storage, etc., may affect the fungicidal potency of these peanut testae.

The biocidal activities of *Dacryodes edulis* and *Garcinia kola* testae have been reported [49]. The antimicrobial activities of these testae are associated with the presence of phytochemicals (alkaloids, saponins, etc.), and was confirmed in experimental studies [50, 51]. The methanolic extract of *Simmondsia chinensis* testa (Link) C.K. Schneid exhibited no fungicidal activities against *Candida albicans* [52], indicating that not every grain testa could inhibit microbial growth.

All the studies mentioned above support the fact that the biological structures of the grains may have the potential to prevent microbial proliferation. These claims cannot be guaranteed when the structures covering the seeds are destroyed during harvesting or drying. Therefore, care should be taken to minimize the destruction of these structures on grains during or after harvest. Busta et al. [53] reported that pathogens lack the enzyme necessary to break down the protective layers covering grains.

2.3 Moisture content (MC)

The oldest method of preserving food is controlling the MC. It is applicable during grain storage since the moisture influences the growth of microorganisms and subsequent production of mycotoxins. The water requirement of microbes is known as the water activity (a_w) of the food or environment and is defined as the ratio of the water vapor pressure of the food substrate to the vapor pressure of pure water at a constant temperature [47]. The a_w of grains describes the degree to which water is bound in the grains, its availability to participate in chemical/biochemical reactions, and its accessibility to facilitate the growth of microorganisms [53] which leads to the synthesis of metabolites.

Cereals have an a_w between 0.10 and 0.20 when adequately dried, making it difficult for microbes to reproduce. Although the optimum MC for growth and subsequent toxin production for the various aflatoxigenic fungi varies, many achieve the best growth and toxin synthesis at an MC of 17.5% [53, 54]. *Aspergillus* requires about 13% moisture or a relative humidity of 65% (a_w, of 0.65) for growth and toxin synthesis [55].

The highest *A. flavus* population was observed at a_w = 0.95. A_w significantly altered the AFB1 produced and the expression of *aflR* at a_w 0.90 and 0.95 respectively. The optimum expression of the *nor-1* gene was at a_w 0.95 and 0.90, whereas deficient expression occurred in the driest treatment (a_w 0.85) [56]. Molds were unable to germinate when the a_w of the grains remained below 0.60. Also, when molds are allow to flourish, they could predispose the stored grain to mite and insect infestation [3, 57] because mites feeds on molds. Co-culturing *A. parasiticus* with *S. lactis* and *Lactobacillus casei* suppressed aflatoxin synthesis [54]. In a similar study, Faraj et al. [58] reported a significant reduction in total aflatoxins synthesized when fungi (*A. niger* and *Rhizopus oryzae*) were co-cultured with a bacterium (*Bacillus stearothermophilus*). Since aflatoxins synthesis was minimal at 40°C and high between 8°C and 40°C, the authors associated the findings to the temperature differential between the strains [59]. However, mycotoxins such as rubratoxins from *Penicillium purpurogenum*, cerulenin from *Cephalosporium caerulens*, and *Acrocylindrium oryzae* inhibited fungi growth at the same time enhance aflatoxin synthesized [45, 60].

The growth of *Trichoderma asperellum* (strains PR10, PR11, PR12, and 659-7) was reported being sensitive to a_w reduction [61]. Therefore, lowering a_w could inhibit the growth of fungi. According to [62], grains stored for a year, 8–9 months, and weeks should have MC about 9%, 13%, and 14%, respectively. A low MC could curb problems like molds infestations, discoloration, respiration loss, insect damage, and moisture absorption.

Adequate drying of grains (produce) to lower moisture levels is critical to create unfavorable conditions to inhibit microbial and insect proliferation. It is recommended to dry harvested produce to safer moisture levels of 10–13%. Low moisture help keep grains longer without losing nutrients and other vital bioactive compounds [63, 64]. Water activity in stored grains could increase depending on climatic conditions, cellular respiration of microorganisms, or urine from rodents. Improper drying, especially during winter or autumn, could also elevate a_w levels.

2.4 pH, acidity and redox potential

For centuries, people have learned to increase the acidity of food either through fermentation, or by adding weak acids in the form of preservatives. These techniques have proven successful. Organic acids are effective preservatives in their undissociated state. pKa is the term used to illustrate the dissociation of an acid. Therefore, lowering the pH of grains increases the effectiveness of organic acids as preservatives [39, 53].

Naturally, grains in the field are undried and possess high pH; however, drying decreases the MC and subsequently the a_w, thereby reducing the pH. Adadi and Obeng [65] reported that the lower the pH value the higher the total acidity (TA), which inhibits the growth of microorganisms. The pH of grains could interact with other parameters (a_w, salt, temperature, redox potential) in the food to inhibit microbial growth. The general rule of food microbiology states that pathogens do not grow, or grow slowly, at pH below 4.6- but there are exceptions. For instance, at pH 4.2, an organism was able to survive and synthesize a mycotoxin [66].

Rice and maize have pH about 6.02 ± 0.01 and 6.53 ± 0.01 during the rainy season and 6.20 ± 0.20 and 6.42 ± 0.12, respectively, in the dry season [67]. The season seems to influence the a_w and the TA, thus altering the pH of the grains. The rainy season is defined by continuous rain, resulting in the elevation of the MC of the grains, which affects the pH. The pH range of beans (string and lima) is between 4.6 and 6.5 [53].

According to [68], fungi can secrete butyrate, oxalate, maleate, citrate, gluconate, and succinate into their environment, thereby changing the acidity of the ecological niche. *Sclerotinia sclerotiorum* and *Botrytis* sp. secrete oxalic acid while *Penicillium* spp., and *Aspergillus* spp., synthesize mainly gluconic and citric acids [69–71]. Fungi can grow comfortably in pH above 8.5; however, below pH 2.2, their growth was inhibited. Microorganisms can modify the pH of the environment in which they reside, making it challenging for farmers to control the pH of stored grain. A phenomenon like this could lead to significant economic loss due to microbial proliferation. The synthesis of ochratoxin A was maximized at lower pH [72]. Different fungi strains (*Trichoderma harzianum*, *Trichoderma aureoviride*, and *Trichoderma viride*) can grow over a broader pH range (from 2.0 to 6.0), with optimal growth at pH = 4.0 [73]. Hence, adjusting the pH is a great way of inhibiting the germination of any fungi spores on stored grains.

The redox potential (Eh) of a substance is the ratio of the total oxidizing (electron-accepting) power to the whole reducing (electron-donating) energy of the material. It is quantified in millivolts (mV) at pH 7.0. Eh correlates to the pH of a substrate [47]. Generally, aerobes, facultative anaerobes, and anaerobes grow well at Eh between +500 to +300 mV, +300 to −100 mV, and + 100 to less than −250 mV, respectively [74]. Some microorganisms require an Eh of less than +60 mV for growth; nevertheless, slower growth rates were observed at higher Eh values [53]. The Eh values of wheat (whole grain), wheat (germ), and barley (ground) is within −320 to −360, −470, and +225, respectively [46]. Oxidants such as $KMnO_4$, $NaClO_4$, or Fe_2O_3 can influence the Eh of a material [75]. The growth of *Fusarium oxysporum* and *Rhizoctonia solani* were suppressed when decomposable organic material was introduced [76, 77]. pH and Eh can impact a wide range of fungal physiological processes (regulation and expression of genes) [78–80] thus complicating the storage process. Therefore, controlling the Eh and pH of grains is necessary to manipulate fungi growth during storage.

2.5 Temperature

All microorganisms have a defined temperature range within which they can grow and synthesize toxins which cause food poisoning. Therefore, understanding the temperatures range, coupled with other intrinsic and extrinsic factors, are crucial to selecte the proper storage conditions for grain storage. Temperature has a dramatic impact on the growth and lag period of an organism. The growth rates of most micro-organisms are favored at low temperatures, though there are exceptions. Reaction rates for specific enzymes in an organism become slower at lower temperatures. Also, low temperatures minimize the fluidity of the cytoplasmic membrane, thus interfering with transport mechanisms in the cell [46, 53]. The expression of proteins are temperature regulated. A slight change in temperature can influence bacterial and archaeal community structure. 16S rRNA genes were altered due to changes in temperature [81, 82]. A wide range of temperatures play a vital role in the growth and synthesis of toxins in fungi. For instance, *Penicillium* and *Cladosporium* were able to grow below 20°C whereas the growth of *Aspergillus* species were inhibited. However, at a temperature above 20°C, the growth was maximized [55]. Virulent *A. niger* has optimal growth between 30–35°C [83], thus, rendering stored produce susceptible to a toxin secreted by these fungi. The growth rates of *Phoma* spp. 1, *Phoma exigua*, *Mortierella gamsii*, and *Mortierella* sp. 1 was high at 4°C [84]. Warmer (33°C) and more humid conditions may increase aflatoxin prevalence. However, the opposite scenario is expected in tropics, since most aflatoxigenic fungi will not survive the expected 40°C [45, 85].

The knowledge of optimal temperature for microbial growth and mycotoxin synthesis gives more accurate assessment of the potential risk to human health [72]. Molds can grow over a broader range of temperatures, from below freezing to temperatures over 50°C. For a given substrate, the rate of mold growth decrease with decreasing temperature and water availability. Below 17°C grains are susceptible to insect infestation; however, mite infestations can occur between 3 and 30°C [86]. Degradation of fungi mycotoxins can occur at 40°C [58]. Therefore, keeping the temperature of the storage room elevated could be of valuable aid in detoxification and probable killing of stored microorganisms.

3. Effects of mycotoxins on human health

Mycotoxins are considered a significant health and economic problem. Mycotoxins can find their way to the human body by way of contaminated food, skin contact, or inhalation [87, 88]. The most common form of exposure is through oral ingestion of contaminated food [89].

The level of exposure and the type of mycotoxins which one is exposed to determine the nature of adverse effects on the human, either in the form of an allergic reaction, infections, or a toxic disease [90]. The seriousness of mycotoxins depends on the toxicity of the mycotoxin involved, the age, wellbeing of the exposed individual, and the length of exposure [91]. Mycotoxicosis is the disease caused by mycotoxins. Mycotoxins such as aflatoxins have been documented causing liver cancer [92]. Other serious conditions, such as chronic interstitial nephropathy, Balkan endemic nephropathy, and urothelial tumors, as well as testicular cancer in men, have also been linked to mycotoxins [93]. Acute diseases, namely abdominal pains, headache, dizziness, throat irritation, and nausea, have also been associated with mycotoxin exposure in humans [94]. It is, therefore, important to ensure that grains are free of mycotoxin contamination.

3.1 Methods of detecting and analyzing mycotoxins

The hazardous effects of mycotoxins on humans and animals had called for the development of rapid methods for their detection and quantification in cereals and other foods. However, sampling methods, extraction, and the instrument used could alter mycotoxin quantification. In response, Rahmani et al. [95] compiled a good comprehensive review to address the challenges mentioned above.

The impact of the sampling on sample preparation and analytical instrument contribute to the total variance during the analysis of ochratoxin A (OTA) in flour and aflatoxinB1 (AFB1) in oats was recently reported. The authors suggested that increasing sample weight (size) could potentially reduce the high heterogeneity encountered [96, 97]. For efficient extraction, methods of detection and quantification of mycotoxins, the reader(s) are referred to the following good sources [95, 98–101].

4. Some conventional methods of controlling grains microbial contamination

Contamination of stored grains by fungi mycotoxins has resulted in economic losses of food products, which could have been used to feed the less privileged (i.e., refugees, natural disaster victims, etc.). Therefore, preservation of grains during storage is necessary to maintain food security. Moreover, with the growing population of the world, more food will be required to feed folks. Some conventional approaches used in preserving grains are listed in **Table 1** besides those described below.

4.1 Organic acids (OA)

High-moisture grains are prone to deterioration during storage if moisture exceeds 14%. For this reason, in the 1970s, chemicals were used to preserve high moisture grains. Propionic acid was used alone (applied worldwide) or in combination with acetic acid, isobutyric acid. Formaldehyde was mostly used in Europe to inhibit the growth of mold and bacteria in outdoor storage of grains. However, when galvanized steel equipment are used to store acid treated grains, extreme corrosion occurred. Thus, lining the bins with oil was recommended. The combinations of propionic acid and sodium benzoate curbed the issue of corrosion, and less harmful compared to pure propionic acid [114–116]. Coating the bins with silver nanoparticle protective paints [117] could prevent corrosion and exert fungicidal activities.

Reference	Methods	Limitations
[4, 102, 103]	Debranning	• Not entirely suitable for wheat due to the crease on the wheat kernels. • Whole-grain demand in the market.
[104–106]	Pesticides	• High environmental impacts. • Direct negative impact on human health. • Increasing resistance against pesticides.
[107–110]	Ozone	• The cost of treatment can be relatively high due to complex technology. • Limited to highly vented packages or open-top containers.
[111–113]	Irradiation	• Can negatively modify the quality and technological properties of cereals and cereal products

Modified with permission from ref. 4496530764014 [122].

Table 1.
Some conventional approaches of grains preservation.

OA can increase moisture content and penetrate the endosperm, thus alter the functionality of the grains [118, 119]. It could also modify the nutritional composition of the stored grain, consequently decreasing the quantity and quality of nutrients. The combination of organic acids, such as propionic, sorbic, and acetic acids, as well as their salts, had antimould activities, which extended the shelf life of bakery products [36]. Similarly, calcium propionate (0.003%), potassium sorbate (0.03%), and sodium benzoate (0.3%) suppressed the growth and mycotoxin production in *Eurotium, Aspergillus* and *Penicillium*. However, the author claimed that a_w and pH contributed to the effectiveness of the compounds and should therefore be carefully considered during application [115]. High sorbate concentration altered the sensorial properties of food [120]; therefore, the concentration used is crucial to maintain grain quality after storage. Propionic acid and its salts exhibited antimicrobial effect against *Bacillus* spp., and was ascribed to their high MW fatty acids [120]. Valerio et al. [121] tested the antifungal activities of organic acids synthesized by lactic acid bacteria (LAB) isolated from a semolina ecosystem. The results showed that all the acids produce by the LAB had inhibitory effects on the test species (*Penicillium roqueforti*, *A. niger*, and *Endomyces fibuligera*). This approach could be classified as biopreservation since the metabolites of living organisms were used to inhibit the growth of microorganisms on the product.

4.2 Drying

According to [122], drying is the phase of postharvest processing during which grains are dried to achieve low MC, thereby guaranteeing safe storage (<0.70 a_w). The MC of adequately dried grains ranged within 10–14%. Russ and coworkers [123] reported that at higher MC, residue of fermentable sugars and other nutrients predispose grains to microbial colonization, resulting in rapid deterioration. Thus, a productive drying process warrants the reduction of moisture, thereby lowering the pH and creating an uninhabitable environment for the germination and proliferation of a microorganism. Dried grains should be allowed to cool before bagging because heat generated during drying could cause a warm spot. Earlier works [36] reported that warm spot in grains support fungal growth, resulting in contamination of grain by mycotoxins. Kumar and coworkers [124] reviewed a paper on heat convection solar drying systems. Some of the techniques described could be employed when drying grains. The low-cost material utilized in manufacturing these dryers, coupled with user friendly, make them ideal for large scale drying, even for small-scale farmers.

Different drying methods have been described: (1) high temperature or heated air-drying; (2) low-temperature air-drying; (3) combined air-drying; (4) dry ration and in-storage cooling method (an alternative to in-dryer cooling) [125, 126].

The expensive nature (cost of power) of artificial drying makes it unpopular, couple with the technicalities involved. For instance, in Russia, sun drying becomes insufficient due to the high MC (i.e., in St Petersburg, Yekaterinburg, etc.); thus, it is impossible to achieve uniform drying of grains. In Africa, sun drying is efficient and effective since there is almost 13-h of sun during the dry season [127]. Applying excessive temperatures (using artificial means) can lead to grains cracking, loss of viability, as well as economic losses [122, 128].

4.3 Chlorine and hypochlorite

Chlorine dioxide (ClO_2) has biocidal activities due to its oxidizing capacity (strong oxidant), and is widely used for decontamination. It is used both in its gaseous and aqueous forms to sanitize food and, exert potent biocidal activity against bacteria, yeasts, and molds [129–133]. All bacteria and their spores in a hospital room were reported killed/inactivate by ClO_2 gas [134].

Poliovirus was found to have been inhibited due to the application of ClO_2, which interreacted with the viral RNA and damaged the genome's ability to act as a template for RNA synthesis [135]. Aqueous ClO_2 was documented to have significantly enhanced the inactivation of *F. graminearum* on wheat at high concentration, (15 mg/L) compared to lower levels (5 and 10 mg/L) [131]. Inexpensive, less corrosive, the ease with which it mixes with air, rapid diffusion, and being easy to use are some merits associated with this method. However, it can produce toxic by-products and interfere with the flavor compounds in the grains. It also requires expensive onsite generation [136–139]. Chlorine solution (0.4%) was ineffective against highly contaminated grains [140, 141]. The reason could be the colonies were mature and had thicker peptidoglycan, hence, the chlorine could not penetrate the cells to reach the genetic material. Another hypothesis could be that the concentration was not enough to destabilize cell and react with the amino acids. Sun and collaborators [133] documented that coupling aqueous sanitizer with gaseous ClO_2 enhanced the decontamination of foodborne and plant pathogens. It also improved the safety, quality, and sensory properties of products (fruits and vegetables). Nevertheless, higher concentrations may cause bleaching or browning.

5. Nanoparticles

The term 'nano' is a Greek word for dwarf, and a nanometer (nm) is 1-billionth of a meter. Nanotechnology has been in existence for decades now, and not an invention of the twentieth century. Nanomaterials and nanoparticles (NPs) are materials that have at least one dimension on the nanoscale (1–100 nm) or whose basic unit in the three-dimensional space is in this range. NPs have a more comprehensive range of applications in food science and technology, drug delivery, biomedical engineering, tissue engineering, textile industry, environment, electronics, agriculture, etc. [10, 142–145]. Nanoparticles are classified as organic (also known as nanocapsules) and inorganic.

Organic NPs act as core shells to shield sensitive bioactive ingredient such as carotenoids [146] against environmental factors, thereby enhancing their bioavailability for safer delivery [10, 147]. Nanoprecipitation, emulsion-diffusion, double emulsification, emulsion-coacervation, polymer coating, etc. are examples of organic NPs [148]. All these techniques are used to prepare the core materials (β-carotene, probiotic bacteria, folic acid, omega fatty acid, protease enzymes, etc.) for encapsulation. Fluorescent organic NPs have recently been used to develop nanosensors [149] which are used to detect contaminants and other foodborne pathogens as well as in bioremediation [150].

Inorganic NPs have attracted the attention of researchers in the last two decades due to their multiple antimicrobial activities (antifungal or antiviral) coupled with the pronouncement from Food Safety Authority that these NPs are safe and do not affect humans/consumers in any way [151–153]. Silver, silica, and titanium dioxide NPs are the main NPs used in the agri-food industries [154].

5.1 Silver nanoparticles (AgNPs)

Several studies have confirmed the potent biocidal effects of silver nanoparticles (AgNPs) towards fungi [155–158]. Due to their peculiar properties (i.e., optical, electrical, and thermal, and biological properties), AgNPs have been used in several applications: as biocidal agents; medical device coatings; optical sensors; in cosmetics; in the food industry (food products); in diagnostics, orthopedics, drug delivery; as anticancer agents and have greatly enhanced the tumor-killing effects of anticancer drugs [158–163]. Healthcare products, such as scaffolding, burn dressings, water purification systems, and medical devices are manufactured using AgNPs [164, 165]. It was reported that 10 μg/mL AgNPs completely inhibited the growth of 10^7 CFU/mL *E. coli* ATCC 8739 cells in liquid medium. The leakage of reducing sugars and proteins forced respiratory chain dehydrogenases into an inactive state, suggesting that AgNPs penetrated the bacterial cell membrane with high efficiency and could therefore be used in the manufacturing of drugs used against bacterial diseases [158]. AgNPs extracted from *Pistacia atlantica* were effective against important clinical pathogens [166]. AgNPs synthesized (green AgNPs) from the leaf of CRCP (medicinal plant) was utilized against multidrug-resistant (MDR) *P. aeruginosa*, *S. aureus* and CoNS isolates (10^6 CFU each) from post-surgical wound infections. 80 mg/mL AgNPs was reported effective against, *S. aureus* and CoNS isolates but had little effects on *P. aeruginosa*. However, 100-120 mg/mL AgNPs completely inhibited *P. aeruginosa* [153]. These findings shows that the concentration of AgNPs utilize is critical therefore should carefully be considered during application.

The fungicidal activities of AgNPs are documented in many studies [13, 152, 160, 167–170]. Six fungal species (*Aspergillus fumigatus*, *Penicillium brevicompactum*, *Cladosporium cladosporoides*, *Mortierella alpina*, *Chaetomium globosum*, and *Stachybotrys chartarum*) isolated from an indoor environment were used to test the antifungal activity of AgNPs. The results revealed that the presence of AgNPs in concentrations of 30–200 mg/L significantly inhibited or decreased the growth of all the fungi species except *Mortierella* species, which were insensitive to the AgNPs but instead metabolized the AgNPs for its own benefit (the presence of AgNPs in agar substrates significantly enhanced *Mortierella* growth rate) [152]. AgNPs and a conventional antifungal agent, Amphotericin B (for a positive test), were tested against *Saccharomyces cerevisiae* (KCTC 7296), *Trichosporon beigelii* (KCTC 7707), and *Candida albicans* (ATCC 90028). The AgNPs exhibited a minimum inhibition concentration (MIC) value of 2 μg/mL, similar to the positive control [155]. AgNPs was found to effectively suppress growth and AFB1 production in *A. parasiticus* (**Figure 1**) [171]. In a similar study, the addition of AgNP HA1N, AgNP HA2N, and AgNP EH resulted in 88.2%, 67.7% and 83.5% reduction of AFB1 synthesized by *A. flavus* [172]. Also, the fungicidal activity of *Capsicum annuum* L. was recently reported [173]. The active ingredient could be isolated and encapsulated in NPs, which may exhibit potent inhibitory activities against storage pest and microorganism.

5.1.1 Mechanistic action of AgNPs biocidal activities

The potent antimicrobial activity of AgNPs has attracted global attention, hence its application in multiple fields (i.e., food industries, medicine, textile industries, etc.). However, the exact mechanistic action is still not clear, because the mechanism depends on the type of microorganism (i.e., bacteria, fungi, etc.) involved and, since different organisms possess different cell structure, the mechanistic action differ. Several researchers have tried to understand the antimicrobial effects of AgNPs using various model microorganisms, e.g., *E. coli* [158, 174, 175], *P. aeruginosa*, *S. aureus* [175], *V. cholera* [174, 176], *S. cerevisiae* [177, 178] and *S. typhi* [174]. Other groups [179, 180] have also worked on fungi. Mitochondrial dysfunction predispose cells for easier penetration by AgNPs via diffusion and endocytosis. The efficiency of

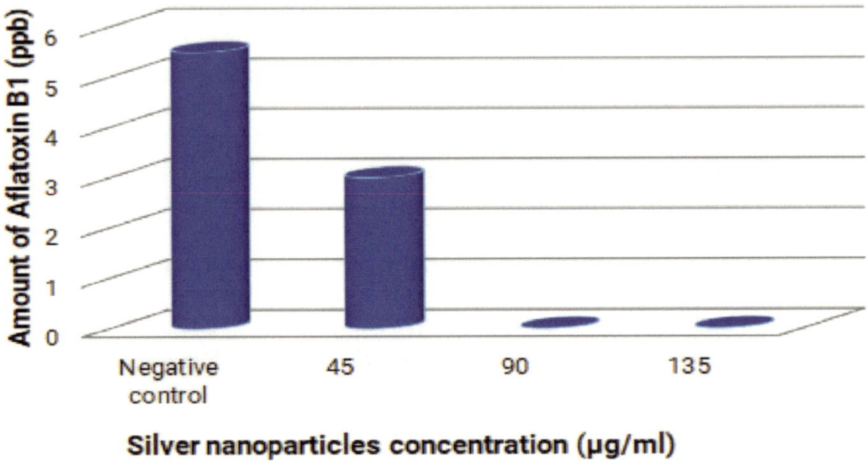

Figure 1.
Inhibition of aflatoxin B_1 production at different concentration of AgNPs. Modified with permission from © Iranian Journal of Medical Sciences [171].

AgNPs uptake by skin keratinocytes depends on the size, shape, pH, zeta potential, and incubation time. Smaller (<5 nm) NPs are more toxic than the larger ones. This could be ascribed to the secure attachment and penetration of the smaller NPs compared to the larger NPs, which requires larger pores to penetrate, into the cell membrane and internalized. AgNPs were able to attach and penetrate cell membrane causing toxicity in *Caenorhabditis elegans*. Ag^0 can interact with molecular oxygen, as well as with other redox-active compounds to produce ionic silver, which then further interact with environmental factors to yield Ag^+ [181–186]. AgNPs ranging from less than 10 nm can inhibit *E. coli* and *P. aeruginosa* due to their potent biocidal activities [187, 188]. Certain viruses were unable to bind to their host cells due to the presence of AgNPs of 1–10 nm, thus starving them to death [189]. Concerning shapes, Pal et al. [190] reported that triangular AgNPs were found to be effective compared to rod and sphere AgNPs. The biocidal efficiency of AgNPs is related to Ag^+, which interact with biological macromolecules (proteins, carbohydrates, nucleic acids, and lipids). When AgNPs adhere to the surface of the cell, it automatically alters membrane properties, undermining the fluidity of the cell. AgNPs can degrade lipopolysaccharide molecules causing them to accumulate inside membrane by forming "pits", thereby increasing membrane permeability [191]. According to reports Ag^+ can inhibit phosphate uptake, resulting in the efflux of phosphate, mannitol, succinate, glutamine, and proline from the cell [192–198].

The minimal bactericidal concentration (MBC) of AgNPs on Gram (+) bacteria was 32 times higher compared to Gram (−) cells [199]. Thus, the sensitivity of the cell wall depends on the class of microorganisms. Research [174] also demonstrated that AgNPs can interact with bacterial cell membranes. Furthermore, the AgNPs found inside the cells are the same sizes as the ones interacting with the membrane, therefore providing more evidence to support the theory that particles that interact with the membrane penetrated into the bacteria.

Several studies [176, 200, 201] have reported that the positive charge of AgNPs is crucial for its antimicrobial activity through the electrostatic attraction with the negatively charged cell membrane of the microorganism.

The permeability of the cell membrane was altered after treatment with AgNPs, resulting in the leaking of reducing sugars and proteins which induced respiratory chain dehydrogenases into inactive state. The amount of reducing sugars leaked after 2 h was 102.5 and 30 µg/mg per bacterial dry weight in the treated and the control cells, respectively. While the activity of respiratory chain dehydrogenases of positive control increased at 37 ± 2, nearly no change was observed in negative control cells. Furthermore, the enzymatic activity of cells treated with 5 μg/mL AgNPs decreased [158]. The survival rate of bacterial species decreased with increase in the adsorption of AgNPs. Additionally, the adsorption and toxicity of AgNPs on *P. aeruginosa*, *M. luteus*, *B. subtilis*, *B. barbaricus*, and *K. pneumonia* was optimum at pH 5, NaCl concentration of <0.5 M. A manifestation of less toxicity was noticed at pH 9 and NaCl concentration >0.5 M, indicating that the environmental pH under which the microorganism grows plays a crucial role in either protecting or exposing it to rapid interaction with the AgNPs [185]. The ability of AgNPs to bind, interact, deform, and induce DNA damage was documented [181, 202–204]. Hackenberg and coworkers [203] used comet assay and chromosomal aberration (CA), a method previously recommended by [205], to determine the damage AgNPs inflict on DNA. In both methods, maximum damage to human mesenchymal stem cells occurred less than an hour after treatment (0.1 μg/mL). Circular dichroism spectra analysis of treated calf thymus DNA revealed that AgNPs interacted and formed a new complex with the double-helical DNA, then induced an alteration of non-planar and change the orientations of DNA bases which act as an intercalator, increasing the stability of DNA which in turn increase the Tm value of the DNA [202]. A researcher [206] suggested that AgNPs can interact with nucleic acids by forming bonds with pyrimidine bases, thus condensing DNA and inhibiting replication. In a recent study, Li et al. [207] showed that citrate-AgNPs (C-AgNP20) induced different cytomorphological alterations and intracellular distributions in cetacean (bottlenose dolphins (*Tursiops truncatus*)) polymorphonuclear cells (cPMNs) and peripheral blood mononuclear cells (cPBMCs). High dose (10 and 50 μg/mL) of C-AgNP20 triggered apoptosis in cPMNs and cPBMCs (induced cytotoxicity). Additionally, the functional activities of cPMNs (phagocytosis and respiratory burst) and cPBMCs (proliferative activity) were negatively altered at sub-lethal dose of 0.1 and 1 μg/mL. AgNPs induced structural damage to cell wall, intracellular proteins (enzymes), and organelles, leading to the disruption or the collapse of metabolic processes, like antioxidant defense mechanisms, thereby inhibiting growth [177, 178].

The cellular oxidative stress in microbes was enhanced by increasing the concentration of Ag (+) ions [206]. Several reports [208–213] have highlighted the potential antiviral, antifungal, and antibacterial activities of AgNPs and was ascribed to its ability to generate enough reactive oxygen species (ROS), free radicals (i.e., hydrogen peroxide (H_2O_2), superoxide anion (O^{2-}), hydroxyl radical ($OH^{\bullet}$), hypochlorous acid (HOCl)) and singlet oxygen. During mitochondrial oxidative phosphorylation, ROS are produced. Moreover, nicotinamide adenine dinucleotide phosphate (NADPH) oxidase catalyzes series of reactions where molecular oxygen (O_2) is reduced to $O^{2\bullet}$. With dismutation and metal-catalyzed Fenton reaction, the $O^{2\bullet}$ is further reduced to H_2O_2 and $OH^{\bullet}$, respectively [214–216]. Apoptosis and cell membrane damage were induced by ROS, leaving the cells incapable of regulating transport through the plasma membrane, resulting in cell death [217–220]. A research group [221], evaluated the effects of ROS against *S. aureus* and *E. coli*. The results showed the inactivation of lactate dehydrogenase and protein denaturation in both test organisms. Membranal damage allowed influx of calcium, thus inducing intracellular calcium overload, further doubling ROS generation and mitochondrial membrane potential variation [222]. The overproduction of ROS was reported to have interfered with ATP synthesis, leading to DNA damage [223]. Free radicals and ROS (an excessive amount) can inflict damage/stress on the mitochondrial membrane, causing necrosis, peroxidation of lipids, proteins, and DNA damage [206, 224, 225]. According to [184, 225], elevated levels of ROS can stress the endoplasmic reticula and deactivate antioxidant enzymes in cells, resulting in genotoxic effects.

It has been discovered that $OH^{\bullet}$, interacted with constituents of DNA, which led to the breakage of DNA single-strands via the formation of 8-hydroxyl-2′-deoxyguanosine (8-OHdG) DNA adduct [226, 227]. In vivo studies have shown that AgNPs influenced the activity of chicken oxidative stress enzymes [228]. AgNP treatment induced a pronounced ROS in *P. aeruginosa* compared to $AgNO_3$. The expression levels of ROS related proteins (PA4133, Hmp, KatA, CcoP2, SodB, CcpA, RibC, EtfA, and PiuC) were specifically regulated after exposure to AgNPs in concentration and time-related modes. Cells treated with $AgNO_3$ did not show any perturbation in intracellular ROS generation at low levels, which supports the existing theory that oxidative stress is triggered solely by AgNPs at their corresponding concentrations [229]. As reported by [220], the biocidal activities of Ag^+ could also be attributed to its interactions with the thiol-related compounds found in the respiratory enzymes of cells, resulting in cell death. A researcher [230] proposed a theory using Ag with cellular energy production. Essential proteins of prokaryotes and eukaryotes located on the cell exterior and interior (mitochondrial organelles), respectively, deactivated after coming in contact with AgNPs. However, the interior components (mitochondrial proteins) required higher concentrations and much smaller AgNPs before they are rendered inactive, because the cellular membrane acted as a diffusion barrier. Moreover, the eukaryotes possessed numerous biological energy conservation system due it extensive mitochondria when compared to the prokaryotes, thereby predisposing the latter cells to AgNP interaction, hampering cell respiration, which led to cell death.

6. Conclusions

It is shown from the above studies that all the mentioned microorganisms, especially the fungi, are involved in grain contamination and subsequent mycotoxin production during storage. Mechanical damage during harvesting or processing served as an easy route via which microorganisms penetrated the endosperms of seeds, and secrete mycotoxins (aflatoxins, etc.) rendering stored grains unsafe for human consumption. The ability of AgNPs to inhibit microbial growth makes them a promising candidate for utilization in storing grains to minimize the economic losses and food poisoning caused by mycotoxins contamination. Moreover, AgNPs inhibited the synthesis of these mycotoxins by switching off molecular pathways via which they are produced, thus guaranteeing the safety of stored grains for consumption. The utilization of AgNPs could enhance shelf-life, maintain the quality and nutritional values of grains. This innovative method is safe and do not pose a threat to the consumer or the environment.

Acknowledgments

This work was supported by the Ministry of Science and Higher Education of Russian Federation, goszadanie no. 2019-1075.

Author details

Daniel Nsengumuremyi[1,2†], Parise Adadi[3†*], Gavers K. Oppong[4,5], Nadezhda V. Barakova[1] and Elena F. Krivoshapkina[2]

1 Department of Food Biotechnology for Plant Origin Products, ITMO University, St. Petersburg, Russian Federation

2 Solution Chemistry of Advanced Materials and Technology (SCAMT) Institute, ITMO University, St. Petersburg, Russian Federation

3 Department of Food Science, University of Otago, Dunedin, New Zealand

4 Department of Plant Sciences, Rothamsted Research, Harpenden, Hertfordshire, United Kingdom

5 School of Biosciences, University of Nottingham, Sutton Bonington Campus, Loughborough, United Kingdom

*Address all correspondence to: parise.adadi@postgrad.otago.ac.nz or pariseadadi@gmail.com

† These authors contributed equally.

References

[1] Harris LJ, Shebuski JR, Danyluk MD, Palumbo MS, Beuchat LR. Nuts, seeds, and cereals. In: Doyle MP, Buchanan RL, editors. Food Microbiology. Washington: American Society for Microbiology Press; 2013. pp. 203-221

[2] Burkepile DE, Parker JD, Woodson CB, et al. Chemically mediated competition between microbes and animals: Microbes as consumers in food webs. Ecology. 2006;**87**:2821-2831

[3] Hocking AD, Pitt JI, Leong SL. Ochratoxin Reduction in Grapes and Dried Vine Fruits. Final Project Report. Australia: FSAID for Dried Fruits Research and Development Council; 2001

[4] Laca A, Mousia Z, Mario D, Webb C, Pandiella SS. Distribution of microbial contamination within cereal grains. Journal of Food Engineering. 2006;**72**(4):332-338

[5] Achaglinkame MA, Opoku N, Amagloh FK. Aflatoxin contamination in cereals and legumes to reconsider usage as complementary food ingredients for Ghanaian infants: A review. Journal of Nutrition & Intermediary Metabolism. 2017;**10**:1-7

[6] Nierop SNEV, Rautenbach M, Axcell BC, Cantrell IC. The impact of microorganisms on barley and malt quality—A Review. Journal of the American Society of Brewing Chemists. 2006;**64**(2):69-78

[7] Roller S, Covill N. The antifungal properties of chitosan in laboratory media and apple juice. International Journal of Food Microbiology. 1999;**47**(1-2):67-77

[8] Sagoo S, Board R, Roller S. Chitosan inhibits growth of spoilage micro-organisms in chilled pork products. Food Microbiology. 2002;**19**(2-3):175-182

[9] Gurunathan S, Han JW, Kwon DN, Kim JH. Enhanced antibacterial and anti-biofilm activities of silver nanoparticles against Gram-negative and Gram-positive bacteria. Nanoscale Research Letters. 2014;**9**(1):373

[10] Dasgupta N, Ranjan S, Chakraborty AR, Ramalingam C, Shanker R, Kumar A. Nanoagriculture and water quality management. In: Ranjan S, Dasgupta N, Lichtfouse E, editors. Nanoscience in Food and Agriculture. Switzerland: Springer International Publishing; 2016. pp. 1-42

[11] Barakova NV, Sharova NY, Juskauskajte AR, Mityukov AS, Romanov VA, Nsengumuremyi D. Fungicidal activity of ultradisperse humic sapropel suspensions. Agronomy Research. 2017;**15**(3):639-648

[12] Barras F, Aussel L, Ezraty B. Silver and antibiotic, new facts to an old story. Antibiotics. 2018;**7**(3):79

[13] Matei P, Martín-Gil J, Iacomi BM, Pérez-Lebeña E, Barrio-Arredondo M, Martín-Ramos P. Silver nanoparticles and polyphenol inclusion compounds composites for *Phytophthora cinnamomi* mycelial growth inhibition. Antibiotics. 2018;**7**(3):76

[14] Ogawa A, Takakura K, Sano K, Kanematsu H, Yamano T, Saishin T, et al. Microbiome analysis of biofilms of silver nanoparticle-dispersed silane-based coated carbon steel using a next-generation sequencing technique. Antibiotics. 2018;**7**(4):91

[15] Sim W, Barnard R, Blaskovich M, Ziora Z. Antimicrobial silver in medicinal and consumer applications: A patent review of the past decade (2007-2017). Antibiotics. 2018;**7**(4):93

[16] Vasil'Kov A, Dovnar R, Smotryn S, Iaskevich N, Naumkin A. Plasmon

resonance of silver nanoparticles as a method of increasing their antibacterial action. Antibiotics. 2018;**7**(3):80

[17] Barros C, Fula S, Stanisic D, Tasic L. Biogenic nanosilver against multidrug-resistant bacteria (MDRB). Antibiotics. 2018;**7**(3):69

[18] RUSSIA—Food and Agriculture Organization. Available from: http://www.fao.org/fileadmin/templates/est/meetings/wto_comm/Trade_Policy_Brief_Russia_final.pdf [Accessed: 28 November 2019]

[19] Lafiandra D, Riccardi G, Shewry PR. Improving cereal grain carbohydrates for diet and health. Journal of Cereal Science. 2014;**59**(3):312-326

[20] Daczkowska-Kozon EG, Bednarczyk A, Biba M, Repich K. Bacteria of bacillus cereus group in cereals at retail. Polish Journal of Food and Nutrition Sciences. 2009;**59**(1):53-59

[21] Huntington GB. Starch utilization by ruminants: From basics to the bunk. Journal of Animal Science. 1997;**75**(3):852

[22] Canadian International Grains Institute. Grains & Oilseeds: Handling, Marketing, Processing. Winnipeg: Canadian International Grains Institute; 1993

[23] Watson CA, Reckling M, Preissel S, Bachinger J, Bergkvist G, Kuhlman T, et al. Grain legume production and use in European agricultural systems. Advances in Agronomy. 2017;**144**:235-303

[24] Scalbert A, Manach C, Morand C, Remesy C, Jimenez L. Dietary polyphenols and the prevention of diseases. Critical Reviews in Food Science and Nutrition. 2005;**45**(4):287-306

[25] Strohle A, Waldmann A, Wolters M, Hahn A. Vegetarian nutrition: Preventive potential and possible risks. Part 1: Plant foods. Wiener Klinische Wochenschrift. 2006;**118**(19-20):580-593

[26] Bassett C, Boye J, Tyler R, Oomah BD. Molecular, functional and processing characteristics of whole pulses and pulse fractions and their emerging food and nutraceutical applications. Food Research International. 2010;**43**(2):397-398

[27] Day L. Proteins from land plants—Potential resources for human nutrition and food security. Trends in Food Science and Technology. 2013;**32**(1):25-42

[28] Dixon RM, Hosking BJ. Nutritional value of grain legumes for ruminants. Nutrition Research Reviews. 1992;**5**(1):19-43

[29] Mikić A, Perić V, Đorđević V, Srebrić M, Mihailović V. Anti-nutritional factors in some grain legumes. Biotechnology in Animal Husbandry. 2009;**25**:1181-1188

[30] Soetan KO, Oyewole OE. The need for adequate processing to reduce the antinutritional factors in plants used as human foods and animal feeds: A review. African Journal of Food Science. 2009;**3**(9):223-232

[31] Osman MA. Effect of different processing methods, on nutrient composition, antinutrional factors, and in vitro protein digestibility of dolichos lablab bean [*Lablab purpuresus* (L) Sweet]. Pakistan Journal of Nutrition. 2007;**6**(4):299-303

[32] Osman MA. Effect of traditional fermentation process on the nutrient and antinutrient contents of pearl millet during preparation of Lohoh. Journal of the Saudi Society of Agricultural Sciences. 2011;**10**(1):1-6

[33] Trushenski JT, Kasper CS, Kohler CC. Challenges and

opportunities in finfish nutrition. North American Journal of Aquaculture. 2006;**68**(2):122-140

[34] Ayadi FY, Rosentrate KA, Muthukumar K. Alternative protein sources for aquaculture feeds. Journal of Aquaculture Feed Science and Nutrition. 2012;**4**(1):1-26

[35] Zhang Y, Øverland M, Sørensen M, Penn M, Mydland LT, Shearer KD, et al. Optimal inclusion of lupin and pea protein concentrates in extruded diets for rainbow trout (*Oncorhynchus mykiss*). Aquaculture. 2012;**344-349**:100-113

[36] Magan N, Aldred D. Managing microbial spoilage in cereals and baking products. In: de Blackburn C. Food Spoilage Microorganisms. Cambridge, UK: Woodhead Publishing Ltd; 2006. pp. 194-212

[37] Alexandraki V, Tsakalidou E, Papadimitriou K, Holzapfel WH. Food Agriculture Organization Report 2013: Status and Trends of the Conservation and Sustainable Use of Microorganisms in Food Processes. Italy: Food Agriculture Organization; 2013

[38] Allsopp D, Seal KJ, Gaylarde CC. Introduction to Biodeterioration. Cambridge: Cambridge University Press; 2004

[39] International Commission on Microbiological Specifications for Foods (ICMSF). Microbial Ecology of Foods. Vol. 2. Food Commodities. New York, NY: Academic Press; 1980

[40] Dhand NK, Joshi DV, Jand SK. Fungal contaminants of dairy feed and their toxigenicity. The Indian Journal of Animal Sciences. 1998;**68**:1095-1096

[41] Torp M, Nirenberg HI. *Fusarium langsethiae* sp. nov. on cereals in Europe. International Journal of Food Microbiology. 2004;**95**(3):247-256

[42] Burger HM, Shephard G, Louw W, Rheeder J, Gelderblom W. The mycotoxin distribution in maize milling fractions under experimental conditions. International Journal of Food Microbiology. 2013;**165**(1):57-64

[43] Liu J, Sun L, Zhang N, Zhang J, Guo J, Li C, et al. Effects of nutrients in substrates of different grains on aflatoxin B1 production by *Aspergillus flavus*. BioMed Research International. 2016:1-10

[44] Li X, Zhao L, Fan Y, Jia Y, Sun L, Ma S, et al. Occurrence of mycotoxins in feed ingredients and complete feeds obtained from the Beijing region of China. Journal of Animal Science and Biotechnology. 2014;**5**(1):37

[45] Makun HA, Dutton MF, Njobeh PB, Gbodi TA, Ogbadu GH. Aflatoxin contamination in foods and feeds: A special focus on Africa. In: Eissa HF, editor. Trends in Vital Food and Control Engineering. London: InTech; 2012

[46] Mossel DAA, Corry JEL, Struijk CB. Essentials of the Microbiology of Foods: A Textbook for Advanced Studies. Chichester: Wiley; 1995

[47] Jay JM. Modern Food Microbiology. 6th ed. Gaithersburg, MD: Aspen; 2000

[48] Xia WX, Zhao P, Wang JB, Li ZJ, Lee NA. Antimicrobial activities of polyphenol extract of peanut skins. In: Proceedings of the 2nd International Conference on Biomedical and Biological Engineering (BBE 2017); 2017

[49] Ebana R, Edet U, Ekanemesang U, Ikon G, Ekpenyong J, Ntukidem N, et al. Comparison of antimicrobial activity and phytochemical screening of seeds and testas of *Dacryodes edulis* and *Garcinia kola*. Journal of Advances in Microbiology. 2016;**1**(3):1-7

[50] Ajibesin KK, Ekpo BAJ, Bala DN. Antimicrobial activities of the

alkaloids fractions of the leaves of *Combretum zenkeri*. Engl. and Diels (Combretaceae). African Journal of Pharmaceutical Research and Development. 2006;**2**:63-66

[51] Okwu D, Nnamdi F. Evaluation of the chemical composition of *Dacryodes edulis* and *Raphia hookeri* Mann and Wendl exudates used in herbal medicine in South Eastern Nigeria. African Journal of Traditional, Complementary, and Alternative Medicines. 2008;**5**(2):194-200

[52] Al-Qizwini H, Al-Khateeb E, Mhaidat NM, Maraqa A. Antioxidant and antimicrobial activities of Jordanian *Simmondsia chinensis* (Link) C.K. Schneid. European Scientific Journal. 2014;**10**(27):229-241

[53] Busta F, Suslow T, Parish M, Beuchat L, Farber J, Garrett E, et al. The use of indicators and surrogate microorganisms for the evaluation of pathogens in fresh and fresh-cut produce. Comprehensive Reviews in Food Science and Food Safety. 2003;**2**(s1):179-185

[54] Ominski KH, Marquardi RR, Sinha RN, Abramson D. Ecological aspects of growth and mycotoxin production by storage fungi. In: Miller JD, Trenholm HL, editors. Mycotoxins in Grains: Compounds Other Than Aflatoxins. St. Paul: Eagan Press; 1994. pp. 287-314

[55] Atanda SA, Pessu PO, Agoda S, Isong IU, Adekalu OA, Echendu MA, et al. Fungi and mycotoxins in stored foods. African Journal of Microbiology Research. 2011;**5**(25):4373-4382

[56] Abdel-Hadi A, Carter D, Magan N. Temporal monitoring of the nor-1 (aflD) gene of *Aspergillus flavus* in relation to aflatoxin B1 production during storage of peanuts under different water activity levels. Journal of Applied Microbiology. 2010;**109**(6):1914-1922

[57] Wicklow DT. The mycology of stored grain: An ecological perspective. In: Jayas DS, White NDG, Muir WE, editors. Stored Grain Ecosystems. New York: Marcel Dekker; 1995. pp. 197-249

[58] Faraj MK, Smith JE, Harran G. Aflatoxin biodegradation: Effects of temperature and microbes. Mycological Research. 1993;**97**(11):1388-1392

[59] Hill RA, Wilson DM, McMillan WW, Widstrom NW, Cole RI, Sanders TH, et al. Ecology of the *Aspergillus flavus* group and aflatoxin formation in maize and groundnut. In: Lacey J, editor. Trichothecenes and Other Mycotoxins. Chichester, UK: Wiley; 1985. pp. 79-86

[60] Smith JE, Moss MO. Mycotoxins. Formation, Analysis and Significance. Chichester/New York: Wiley; 1985

[61] Begoude B, Lahlali R, Friel D, Tondje P, Jijakli M. Response surface methodology study of the combined effects of temperature, pH, and aw on the growth rate of *Trichoderma asperellum*. Journal of Applied Microbiology. 2007;**103**(4):845-854

[62] Rice Knowledge Bank. Moisture Content for Safe Storage [Internet]. Available from: http://www.knowledgebank.irri.org/step-by-step-production/postharvest/storage/moisture-content-for-safe-storage [Accessed: 28 November 2019]

[63] Lanyasunya TL, Wamae LW, Musa HH, Olowofeso O, Lokwaleput IK. The risk of mycotoxins contamination of dairy feed and milk on smallholder dairy farms in Kenya. Pakistan Journal of Nutrition. 2005;**4**(3):162-169

[64] Atanda O, Makun HA, Ogara IM, Edema M, Idahor KO, Eshiett ME, et al. Fungal and mycotoxin contamination

of Nigerian foods and feeds. In: Makun HA, editor. Mycotoxin and Food Safety in Developing Countries. Rijeka: InTech; 2013

[65] Adadi P, Obeng AK. Assessment of bacterial quality of honey produced in Tamale metropolis (Ghana). Journal of Food and Drug Analysis. 2017;**25**(2):369-373

[66] Smelt JPPM, Raatjes GJM, Crowther JS, Verrips CT. Growth and toxin formation by *Clostridium botulinum* at low pH values. Journal of Applied Bacteriology. 1982;**52**(1):75-82

[67] Tawo EN, Abara AE, Malu SP, Alobi N. Evaluation of pH levels in some common carbohydrate food items consumed by communities in the Central Senatorial District of Cross River State, South-South of Nigeria. Pakistan Journal of Nutrition. 2009;**8**(9):1387-1390

[68] Vylkova S. Environmental pH modulation by pathogenic fungi as a strategy to conquer the host. PLoS Pathogens. 2017;**13**(2):e1006149

[69] Manteau S, Abouna S, Lambert B, Legendre L. Differential regulation by ambient pH of putative virulence factor secretion by the phytopathogenic fungus *Botrytis cinerea*. FEMS Microbiology Ecology. 2003;**43**(3):359-366

[70] Ruijter GJ, Visser J. Characterization of *Aspergillus niger* phosphoglucose isomerase. Use for quantitative determination of erythrose 4-phosphate. Biochimie. 1999;**81**(3):267-272

[71] Prusky D, Yakoby N. Pathogenic fungi: Leading or led by ambient pH? Molecular Plant Pathology. 2003;**4**(6):09-516

[72] Passamani FRF, Hernandes T, Lopes NA, Bastos SC, Santiago WD, Cardoso MDG, et al. Effect of temperature, water activity, and pH on growth and production of Ochratoxin A by *Aspergillus niger* and *Aspergillus carbonarius* from Brazilian grapes. Journal of Food Protection. 2014;**77**(11):1947-1952

[73] Kredics L, Manczinger L, Antal Z, Penzes Z, Szekeres A, Kevei F, et al. In vitro water activity and pH dependence of mycelial growth and extracellular enzyme activities of *Trichoderma* strains with biocontrol potential. Journal of Applied Microbiology. 2004;**96**(3):491-498

[74] Ray B. Factors influencing microbial growth in food. In: Fundamental Food Microbiology. 3rd ed. Boca Raton: CRC Press; 2004. pp. 67-79

[75] Savich VI, Kaurichev IS, Draman K. Amendments for regulating the oxidation-reduction potential of soils. Izvestiya Timiryazevskoi Sel'skokhozyaistvennoi Akademii. 1980;**3**:75-82

[76] Blok WJ, Lamers JG, Termorshuizen AJ, Bollen GJ. Control of soilborne plant pathogens by incorporating fresh organic amendments followed by tarping. Phytopathology. 2000;**90**:253-259

[77] Shinmura A. Principle and effect of soil sterilization methods by reducing the redox potential of soil. In: PSJ Soil borne Disease Workshop Report, Vol. 22; 2004. pp. 2-12

[78] Daval S, Lebreton L, Gracianne C, Guillerm-Erckelboudt AY, Boutin M, Marchi M, et al. Strain-specific variation in a soilborne phytopathogenic fungus for the expression of genes involved in pH signal transduction pathway, pathogenesis and saprophytic survival in response to environmental pH changes. Fungal Genetics and Biology. 2013;**61**:80-89

[79] Lehmann S, Serrano M, L'Haridon F, Tjamos SE, Metraux JP. Reactive oxygen species and plant resistance to fungal pathogens. Phytochemistry. 2015;**112**:54-62

[80] Jwa NS, Hwang BK. Convergent evolution of pathogen effectors toward reactive oxygen species signaling networks in plants. Frontiers in Plant Science. 2017;**8**:1687

[81] Norris TB, Wraith JM, Castenholz RW, Mcdermott TR. Soil microbial community structure across a thermal gradient following a geothermal heating event. Applied and Environmental Microbiology. 2002;**68**(12):6300-6309

[82] Stres B, Danevcic T, Pal L, Fuka MM, Resman L, Leskovec S, et al. Influence of temperature and soil water content on bacterial, archaeal and denitrifying microbial communities in drained fen grassland soil microcosms. FEMS Microbiology Ecology. 2008;**66**(1):110-122

[83] Leong SL, Hocking AD, Scott ES. Effect of temperature and water activity on growth and ochratoxin A production by Australian *Aspergillus carbonarius* and *A. niger* isolates on a simulated grape juice medium. International Journal of Food Microbiology. 2006;**110**(3):209-216

[84] Kerry E. Effects of temperature on growth rates of fungi from subantarctic Macquarie Island and Casey, Antarctica. Polar Biology. 1990;**10**(4):293-299

[85] Paterson RRM, Lima N. Toxicology of mycotoxins. In: Lunch AA, editor. Molecular Clinical and Environmental Toxicology. Basel: Springer; 2010. pp. 31-63

[86] FAO, Agricultural Services Bulletin. Grain Storage Techniques: Evolution and Trends in Developing Countries. Proctor DL, editor [Internet]. 1994. Available from: http://www.fao.org/3/T1838E/T1838E00.htm [Accessed: 28 November 2019]

[87] Mayer S, Engelhart S, Kolk A, Blome H. The significance of mycotoxins in the framework of assessing workplace related risks. Mycotoxin Research. 2008;**24**(3):151-164

[88] Bryden WL. Mycotoxins in the food chain: Human health implications. Asia Pacific Journal of Clinical Nutrition. 2007;**16**(S1):95-101

[89] Marin S, Ramos A, Cano-Sancho G, Sanchis V. Mycotoxins: Occurrence, toxicology, and exposure assessment. Food and Chemical Toxicology. 2013;**60**:218-237

[90] Bennett JW, Klich M. Mycotoxins. Clinical Microbiology Reviews. 2003;**16**(3):497-516. DOI: 10.1128/CMR.16.3.497-516.2003

[91] Marín S, Cano-Sancho G, Sanchis V, Ramos AJ. The role of mycotoxins in the human exposome: Application of mycotoxin biomarkers in exposome-health studies. Food and Chemical Toxicology. 2018;**121**:504-518

[92] Liu Y, Wu F. Global burden of aflatoxin-induced hepatocellular carcinoma: A risk assessment. Environmental Health Perspectives. 2010;**118**(6):818-824

[93] Bayman P, Baker JL. Ochratoxins: A global perspective. Mycopathologia. 2006;**162**(3):215-223

[94] Vidal A, Mengelers M, Yang S, Saeger SD, Boevre MD. Mycotoxin biomarkers of exposure: A comprehensive review. Comprehensive Reviews in Food Science and Food Safety. 2018;**17**(5):1127-1155

[95] Rahmani A, Jinap S, Soleimany F. Qualitative and quantitative analysis of

mycotoxins. Comprehensive Reviews in Food Science and Food Safety. 2009;**8**(3):202-251

[96] Yogendrarajah P, Poucke CV, Meulenaer BD, Saeger SD. Development and validation of a QuEChERS based liquid chromatography tandem mass spectrometry method for the determination of multiple mycotoxins in spices. Journal of Chromatography. A. 2013;**1297**:1-11

[97] Skendi A, Irakli MN, Papageorgiou MD. Optimized and validated high-performance liquid chromatography method for the determination of deoxynivalenol and aflatoxins in cereals. Journal of Separation Science. 2016;**39**(8):1425-1432

[98] Pereira VL, Fernandes JO, Cunha SC. Mycotoxins in cereals and related foodstuffs: A review on occurrence and recent methods of analysis. Trends in Food Science and Technology. 2014;**36**(2):96-136

[99] Breidbach A. A greener, quick and comprehensive extraction approach for LC-MS of multiple mycotoxins. Toxins. 2017;**9**(3):91

[100] Bueno D, Munoz R, Marty JL. Common methods to detect mycotoxins: A review with particular emphasis on electrochemical detection. In: Kalcher K, Metelka R, Svancara I, Vytřas K, editors. Sensing in Electroanalysis, 8, Univerzita Pardubice. Pardubice, Czech Republic: University Press Centre; 2013. pp. 85-114

[101] Cao X, Li X, Li J, Niu Y, Shi L, Fang Z, et al. Quantitative determination of carcinogenic mycotoxins in human and animal biological matrices and animal-derived foods using multi-mycotoxin and analyte-specific high-performance liquid chromatography-tandem mass spectrometric methods. Journal of Chromatography B. 2018;**1073**:191-200

[102] Dexter J, Wood P. Recent applications of debranning of wheat before milling. Trends in Food Science and Technology. 1996;7(2):35-41

[103] Pandiella S, Mousia Z, Laca A, Díaz M, Webb C. Debranning technology to improve cereal-based foods. In: Using Cereal Science and Technology for the Benefit of Consumers. Cambridge, England: Woodhead Publishing Limited; 2005. pp. 241-244

[104] Liu Y, Pan X, Li J. A 1961-2010 record of fertilizer use, pesticide application and cereal yields: A review. Agronomy for Sustainable Development. 2014;**35**(1):83-93

[105] Jess S, Kildea S, Moody A, Rennick G, Murchie AK, Cooke LR. European union policy on pesticides: Implications for agriculture in Ireland. Pest Management Science. 2014;**70**(11):1646-1654

[106] Aktar W, Sengupta D, Chowdhury A. Impact of pesticides use in agriculture: Their benefits and hazards. Interdisciplinary Toxicology. 2009;**2**(1):1-12

[107] Palou L, Smilanick JL, Crisosto CH, Mansour M, Plaza P. Ozone gas penetration and control of the sporulation of *Penicillium digitatum* and *Penicillium italicum* within commercial packages of oranges during cold storage. Crop Protection. 2003;**22**(9):1131-1134

[108] Greene AK, Guzel-Seydim ZB, Seydim AC. Chemical and physical properties of ozone. In: O'Donnell C, Tiwary BK, Cullen PJ, Rice RG, editors. Ozone in Food Processing. Chichester, UK: Blackwell Publishing Ltd; 2012. pp. 26-28

[109] Environmental Protection Agency. Wastewater Technology Fact Sheet: Ozone Disinfection [Internet]. 1999.

Available from: https://www3.epa.gov/npdes/pubs/ozon.pdf [Accessed: 29 November 2019]

[110] Sun C, Ji J, Wu S, Sun C, Pi F, Zhang Y, et al. Saturated aqueous ozone degradation of deoxynivalenol and its application in contaminated grains. Food Control. 2016;**69**:185-190

[111] Lung HM, Cheng YC, Chang YH, Huang HW, Yang BB, Wang CY. Microbial decontamination of food by electron beam irradiation. Trends in Food Science and Technology. 2015;**44**(1):66-78

[112] Farkas J, Ehlermann DAE, Mohácsi-Farkas C. Food technologies: Food irradiation. In: Motarjemi Y, Moy G, Todd E, editors. Encyclopedia of Food Safety. Cambridge, Massachusetts: Academic Press; 2014. pp. 178-186

[113] Lynch MF, Tauxe RV, Hedberg CW. The growing burden of foodborne outbreaks due to contaminated fresh produce: Risks and opportunities. Epidemiology and Infection. 2009;**137**(3):307-315

[114] Sauer DB, Burroughs R. Efficacy of various chemicals as grain mold inhibitors. Transactions of the American Society of Agricultural and Biological Engineers. 1993;**17**(3):0557-0559

[115] Lückstädt C. Long Term Preservation of High Moisture Grain and Maize [Internet]. 2010. Available from: http://www.aquafeed.com/docs/2010grapas/GRAPAS 2010_Luckstadt.pdf [Accessed: 29 November 2019]

[116] Kumar A, Vemula PK, Ajayan PM, John G. Silver-nanoparticle-embedded antimicrobial paints based on vegetable oil. Nature Materials. 2008;7(3):236-241

[117] Sabillon L, Stratton J, Rose D, Bianchini A. Effect of saline organic acid solutions applied during soft wheat tempering on microbial load and flour functionality. Cereal Chemistry. 2017;**94**:1-21

[118] Sabillon L, Stratton J, Rose DJ, Flores RA, Bianchini A. Reduction in microbial load of wheat by tempering with organic acid and saline solutions. Cereal Chemistry. 2016;**93**(6):638-646

[119] Olmez H, Kretzschmar U. Potential alternative disinfection methods for organic fresh-cut industry for minimizing water consumption and environmental impact. LWT - Food Science and Technology. 2009;**42**(3):686-693

[120] Khan S, Hashmi S, Saleem Q. Microbial spoilage of bakery products and its control by preservatives. Shodhankan. 2013;**2**:169-177. Available from: https://www.researchgate.net/publication/259495423_Microbial_Spoilage_of_Bakery_Products_and_Its_Control_by_Preservatives [Accessed: 29 November 2019]

[121] Valerio F, Favilla M, Bellis PD, Sisto A, Candia SD, Lavermicocca P. Antifungal activity of strains of lactic acid bacteria isolated from a semolina ecosystem against *Penicillium roqueforti*, *Aspergillus niger* and *Endomyces fibuliger* contaminating bakery products. Systematic and Applied Microbiology. 2009;**32**(6):438-448

[122] Los A, Ziuzina D, Bourke P. Current and future technologies for microbiological decontamination of cereal grains. Journal of Food Science. 2018;**83**(6):1484-1493

[123] Russ W, Mörtel H, Meyer-Pittroff R. Application of spent grains to increase porosity in bricks. Construction and Building Materials. 2005;**19**(2):117-126

[124] Kumar Y. Review on heat convection solar drying in dryer. International Journal of Pure & Applied Bioscience. 2018;**6**(2):1323-1330

[125] Friesen OH. Heated-Air Grain Dryers. Ottawa: Agriculture Canada; 1981

[126] Friesen OH, Huminicki DN. Grain Aeration and Unheated-Air Drying. Manitoba Agriculture: Agdex. 1986. 732-1. Available from: ftp://ftp.ufv.br/dea/Disciplinas/Evandro/Eng671/Aulas/Aula09-2-aeration.pdf [Accessed: 02 November 2019]

[127] Adelaja AO, Babatope BI. Analysis and testing of a natural convection solar dryer for the tropics. Journal of Energy. 2013:1-8

[128] Yaciuk G. Agricultural applications of solar energy. In: Solar Energy Conversion II, Selected Lectures from the 1980 International Symposium on Solar Energy Utilization. Oxford: Pergamon Press; 1980. pp. 337-353

[129] Yang M, Liu J, Zhang X, Richardson SD. Comparative toxicity of chlorinated saline and freshwater wastewater effluents to marine organisms. Environmental Science & Technology. 2015;**49**(24):14475-14483

[130] Haute SV, Tryland I, Escudero C, Vanneste M, Sampers I. Chlorine dioxide as water disinfectant during fresh-cut iceberg lettuce washing: Disinfectant demand, disinfection efficiency, and chlorite formation. LWT. 2017;**75**:301-304

[131] Sun C, Zhu P, Ji J, Sun J, Tang L, Pi F, et al. Role of aqueous chlorine dioxide in controlling the growth of *Fusarium graminearum* and its application on contaminated wheat. LWT. 2017;**84**:555-561

[132] Sun X, Baldwin E, Plotto A, Narciso J, Ference C, Ritenour M, et al. Controlled-release of chlorine dioxide in a perforated packaging system to extend the storage life and improve the safety of grape tomatoes. Journal of Visualized Experiments. 2017;**7**:122

[133] Sun X, Baldwin E, Bai J. Applications of gaseous chlorine dioxide on postharvest handling and storage of fruits and vegetables—A review. Food Control. 2019;**95**:18-26

[134] Lowe JJ, Hewlett AL, Iwen PC, Smith PW, Gibbs SG. Evaluation of ambulance decontamination using gaseous chlorine dioxide. Prehospital Emergency Care. 2013;**17**(3):401-408

[135] Alvarez ME, Obrien RT. Mechanisms of inactivation of poliovirus by chlorine dioxide and iodine. Applied and Environmental Microbiology. 1982;**44**:1064-1071

[136] Richardson SD, Thruston AD Jr, Caughran TV, Collette TW, Patterson KS, Lykins WW Jr. Chemical by-products of chlorine and alternative disinfectants. Food Technololgy. 1998;**52**:58-61

[137] Virto R, Manas P, Alvarez I, Condon S, Raso J. Membrane damage and microbial inactivation by chlorine in the absence and presence of a chlorine-demanding substrate. Applied and Environmental Microbiology. 2005;**71**(9):5022-5028

[138] Sun X, Bai J, Ference C, Wang Z, Zhang Y, Narciso J, et al. Antimicrobial activity of controlled-release chlorine dioxide gas on fresh blueberries. Journal of Food Protection. 2014;**77**(7):1127-1132

[139] Los A, Ziuzina D, Boehm D, Cullen PJ, Bourke P. The potential of atmospheric air cold plasma for control of bacterial contaminants relevant to cereal grain production. Innovative Food Science & Emerging Technologies. 2017;**44**:36-45

[140] Andrews S, Pardoel D, Harun A, Treloar T. Chlorine inactivation of fungal spores on cereal grains. International Journal of Food Microbiology. 1997;**35**(2):153-162

[141] Delaquis P, Bach S. Resistance and sublethal damage. Produce contamination. In: Gomez-Lopez VM, editor. Decontamination of Fresh and Minimally Processed Produce. New Jersey: Wiley-Blackwell Publishing; 2012. pp. 77-86

[142] Laurent S, Forge D, Port M, Roch A, Robic C, Vander EL, et al. Magnetic iron oxide nanoparticles: Synthesis, stabilization, vectorization, physicochemical characterizations, and biological applications. Chemical Revolution. 2010;**110**:2574-2574

[143] Kingsley JD, Ranjan S, Dasgupta N, Saha P. Nanotechnology for tissue engineering: Need, techniques and applications. Journal of Pharmacy Research. 2013;**7**(2):200-204

[144] Ranjan S, Dasgupta N, Chakraborty AR, Samuel SM, Ramalingam C, Shanker R, et al. Nanoscience and nanotechnologies in food industries: Opportunities and research trends. Journal of Nanoparticle Research. 2014;**16**(6):2464

[145] Bartolucci C. Nanotechnologies for agriculture and foods: Past and future. In: Monique AV, Axelos AV, Van de Voorde M, editors. Nanotechnology in Agriculture and Food Science. Weinheim: Wiley; 2017. pp. 3-14

[146] Adadi P, Barakova VN, Krivoshapkina EF. Selected methods of extracting carotenoids, characterization, and health concerns: A review. Journal of Agricultural and Food Chemistry. 2018;**66**:5925-5947

[147] Anton N, Benoit JP, Saulnier P. Design and production of nanoparticles formulated from nano-emulsion templates—A review. Journal of Controlled Release. 2008;**128**:185-199

[148] Mora-Huertas C, Fessi H, Elaissari A. Polymer-based nanocapsules for drug delivery. International Journal of Pharmaceutics. 2010;**385**(1-2):113-142

[149] Zhang X, Zhang X, Yang B, Zhang Y, Wei YA. New class of red fluorescent organic nanoparticles: Noncovalent fabrication and cell imaging applications. ACS Applied Materials & Interfaces. 2014;**6**(5):3600-3606

[150] Zhang X, Zhang X, Yang B, Hui J, Liu M, Chi Z, et al. Facile preparation and cell imaging applications of fluorescent organic nanoparticles that combine AIE dye and ring-opening polymerization. Polymer Chemistry. 2014;**5**(2):318-322

[151] Abdel-Mohsen A, Hrdina R, Burgert L, Abdel-Rahman RM, Hašová M, Šmejkalová D, et al. Antibacterial activity and cell viability of hyaluronan fiber with silver nanoparticles. Carbohydrate Polymers. 2013;**92**(2):1177-1187

[152] Ogar A, Tylko G, Turnau K. Antifungal properties of silver nanoparticles against indoor mould growth. Science of the Total Environment. 2015;**521-522**:305-314

[153] Kasithevar M, Periakaruppan P, Muthupandian S, Mohan M. Antibacterial efficacy of silver nanoparticles against multi-drug resistant clinical isolates from post-surgical wound infections. Microbial Pathogenesis. 2017;**107**:327-334

[154] Higashisaka K, Yoshioka Y, Tsutsumi Y. Applications and safety of nanomaterials used in the food industry. Food Safety. 2015;**3**(2):39-47

[155] Klasen HA. Historical review of the use of silver in the treatment of burns. II. Renewed interest for silver. Burns. 2000;**26**(2):131-138

[156] Silver S. Bacterial silver resistance: Molecular biology and uses

and misuses of silver compounds. FEMS Microbiology Reviews. 2003;**27**(2-3):341-353

[157] Lok CN, Ho CM, Chen R, He Q-Y, Yu WY, Sun H, et al. Proteomic analysis of the mode of antibacterial action of silver nanoparticles. Journal of Proteome Research. 2006;**5**(4):916-924

[158] Li WR, Xie X-B, Shi QS, Zeng H-Y, Ou-Yang Y-S, Chen Y-B. Antibacterial activity and mechanism of silver nanoparticles on *Escherichia coli*. Applied Microbiology and Biotechnology. 2009;**85**(4):1115-1122

[159] Long TC, Tajuba J, Sama P, Saleh N, Swartz C, Parker J, et al. Nanosize titanium dioxide stimulates reactive oxygen species in brain microglia and damages neurons in vitro. Environmental Health Perspectives. 2007;**115**(11):1631-1637

[160] Mukherjee P, Ahmad A, Mandal D, Senapati S, Sainkar SR, Khan MI, et al. Fungus-mediated synthesis of silver nanoparticles and their immobilization in the mycelial matrix: A novel biological approach to nanoparticle synthesis. Nano Letters. 2001;**1**(10):515-519

[161] Chernousova S, Epple M. Silver as antibacterial agent: Ion, nanoparticle, and metal. Angewandte Chemie International Edition. 2012;**52**(6):1636-1653

[162] Gurunathan S, Park JH, Han JW, Kim J-H. Comparative assessment of the apoptotic potential of silver nanoparticles synthesized by *Bacillus tequilensis* and *Calocybe indica* in MDA-MB-231 human breast cancer cells: Targeting p53 for anticancer therapy. International Journal of Nanomedicine. 2015;**10**:4203-4223

[163] Thomas V, Yallapu MM, Sreedhar B, Bajpai SA. Versatile strategy to fabricate hydrogel–silver nanocomposites and investigation of their antimicrobial activity. Journal of Colloid and Interface Science. 2007;**315**(1):389-395

[164] Kim S, Kim H-J. Anti-bacterial performance of colloidal silver-treated laminate wood flooring. International Biodeterioration & Biodegradation. 2006;**57**(3):155-162

[165] Sadeghi B, Rostami A, Momeni S. Facile green synthesis of silver nanoparticles using seed aqueous extract of *Pistacia atlantica* and its antibacterial activity. Spectrochimica Acta Part A: Molecular and Biomolecular Spectroscopy. 2015;**134**:326-332

[166] Kim K-J, Sung WS, Suh BK, Moon S-K, Choi J-S, Kim JG, et al. Antifungal activity and mode of action of silver nano-particles on *Candida albicans*. BioMetals. 2008;**22**(2):235-242

[167] Chladek G, Mertas A, Barszczewska-Rybarek I, Nalewajek T, Żmudzki J, Król W, et al. Antifungal activity of denture soft lining material modified by silver nanoparticles—A pilot study. International Journal of Molecular Sciences. 2011;**12**(7):4735-4744

[168] Qasim M, Singh BR, Naqvi AH, Paik P, Das D. Silver nanoparticles embedded mesoporous SiO_2 nanosphere: An effective anticandidal agent against *Candida albicans* 077. Nanotechnology. 2015;**26**(28):285102

[169] Musa SF, Yeat TS, Kamal LZM, Tabana YM, Ahmed MA, Ouweini AE, et al. Pleurotussajor-cajucan be used to synthesize silver nanoparticles with antifungal activity against *Candida albicans*. Journal of the Science of Food and Agriculture. 2017;**98**(3):1197-1207

[170] Mousavi SAA, Pourtalebi S. Inhibitory effects of silver nanoparticles on growth and Aflatoxin B1 production

by *Aspergillus parasiticus*. Iranian Journal of Medical Sciences. 2015;**40**:501-506

[171] Deabes MM, Khalil WKB, Attallah AG, El-Desouky TA, Naguib KM. Impact of silver nanoparticles on gene expression in *Aspergillus flavus* producer aflatoxin B1. Open Access Macedonian Journal of Medical Sciences. 2018;**6**(4):600-605

[172] Zhao J, Wang L, Xu D, Lu Z. Involvement of ROS in nanosilver-caused suppression of aflatoxin production from *Aspergillus flavus*. RSC Advances. 2017;**7**(37):23021-23026

[173] Go SM, Park MR, Kim HS, Choi WS, Jeong RD. Antifungal effect of non-thermal atmospheric plasma and its application for control of postharvest *Fusarium oxysporum* decay of paprika. Food Control. 2019;**98**:245-252

[174] Morones JR, Elechiguerra JL, Camacho A, Holt K, Kouri JB, Ramırez JT, et al. The bactericidal effect of silver nanoparticles. Nanotechnology. 2005;**16**:2346-2353

[175] Jung WK, Koo HC, Kim KW, Shin S, Kim SH, Park YH. Antibacterial activity and mechanism of action of the silver ion in *Staphylococcus aureus* and *Escherichia coli*. Applied and Environmental Microbiology. 2008;**74**(7):2171-2178

[176] Dibrov P, Dzioba J, Gosink KK, Hase CC. Chemiosmotic mechanism of antimicrobial activity of ag in *Vibrio cholerae*. Antimicrobial Agents and Chemotherapy. 2002;**46**(8):2668-2670

[177] Saulou C, Jamme F, Maranges C, Fourquaux I, Despax B, Raynaud P, et al. Synchrotron FTIR microspectroscopy of the yeast *Saccharomyces cerevisiae* after exposure to plasma-deposited nanosilver-containing coating. Analytical and Bioanalytical Chemistry. 2010;**396**(4):1441-1450

[178] Despax B, Saulou C, Raynaud P, Datas L, Mercier-Bonin M. Transmission electron microscopy for elucidating the impact of silver-based treatments (ionic silver versus nanosilver-containing coating) on the model yeast *Saccharomyces cerevisiae*. Nanotechnology. 2011;**22**(17):175101

[179] Alananbeh KM, Refaee WJA, Qodah ZA. Antifungal effect of silver nanoparticles on selected fungi isolated from raw and waste water. Indian Journal of Pharmaceutical Sciences. 2017;**79**(4):559-567

[180] Majeed S. Biosynthesis and characterization of nanosilver from *Alternata alternaria* and its antifungal and antibacterial activity in combination with fluconazole and gatifloxacin. Biomedical and Pharmacology Journal. 2017;**10**(4):1709-1714

[181] Choi O, Hu Z. Size dependent and reactive oxygen species related nanosilver toxicity to nitrifying bacteria. Environmental Science & Technology. 2008;**42**(12):4583-4588

[182] Meyer JN, Lord CA, Yang XY, Turner EA, Badireddy AR, Marinakos SM, et al. Intracellular uptake and associated toxicity of silver nanoparticles in *Caenorhabditis elegans*. Aquatic Toxicology. 2010;**100**(2):140-150

[183] Lu W, Senapati D, Wang S, Tovmachenko O, Singh AK, Yu H, et al. Effect of surface coating on the toxicity of silver nanomaterials on human skin keratinocytes. Chemical Physics Letters. 2010;**487**(1-3):92-96

[184] Mcshan D, Ray PC, Yu H. Molecular toxicity mechanism of nanosilver. Journal of Food and Drug Analysis. 2014;**22**(1):116-127

[185] Khan SS, Mukherjee A, Chandrasekaran N. Studies on interaction of colloidal silver

nanoparticles (SNPs) with five different bacterial species. Colloids and Surfaces B: Biointerfaces. 2011;**87**(1):129-138

[186] Khan SS, Srivatsan P, Vaishnavi N, Mukherjee A, Chandrasekaran N. Interaction of silver nanoparticles (SNPs) with bacterial extracellular proteins (ECPs) and its adsorption isotherms and kinetics. Journal of Hazardous Materials. 2011;**192**:299-306

[187] Xu X-HN, Brownlow WJ, Kyriacou SV, Wan Q, Viola JJ. Real-time probing of membrane transport in living microbial cells using single nanoparticle optics and living cell imaging. Biochemistry. 2004;**43**(32):10400-10413

[188] Gogoi SK, Gopinath P, Paul A, Ramesh A, Ghosh SS, Chattopadhyay A. Green fluorescent protein-expressing *Escherichia coli* as a model system for investigating the antimicrobial activities of silver nanoparticles. Langmuir. 2006;**22**(22):9322-9328

[189] Elichiguerra JL, Burt JL, Morones JR, Camacho-Bragado A, Gao X, Lara HH, et al. Interaction of silver nanoparticles with HIV-1. Journal of Nanobiotechnology. 2005;**3**:6

[190] Pal S, Tak YK, Song JM. Does the antibacterial activity of silver nanoparticles depend on the shape of the nanoparticle? A study of the gram-negative bacterium *Escherichia coli*. Applied and Environmental Microbiology. 2007;**73**(6):1712-1720

[191] Sondi I, Salopek-Sondi B. Silver nanoparticles as antimicrobial agent: A case study on *E. coli* as a model for Gram-negative bacteria. Journal of Colloid and Interface Science. 2004;**275**(1):177-182

[192] Wijnhoven SW, Peijnenburg WJ, Herberts CA, Hagens WI, Oomen AG, Heugens EH, et al. Nano-silver—A review of available data and knowledge gaps in human and environmental risk assessment. Nanotoxicology. 2009;**3**(2):109-138

[193] Sanford J, Venkatapathy R. State of the science literature review: Everything nanosilver and more. In: Varner K, editor. Scientific, Technical, Research, Engineering, and Modeling Support Final Report. Washington, DC: US Environmental Protection Agency, Office of Research and Development; 2010. pp. 1-197

[194] Johnston HJ, Hutchison G, Christensen FM, Peters S, Hankin S, Stone VA. Review of the in vivo and in vitro toxicity of silver and gold particulates: Particle attributes and biological mechanisms responsible for the observed toxicity. Critical Reviews in Toxicology. 2010;**40**(4):328-346

[195] Kruszewski M, Brzoska K, Brunborg G, Asare N, Refsnes M. Toxicity of silver nanomaterials in higher eukaryotes. In: Fishbein JC, editor. Advances in Molecular Toxicology. Amsterdam: Elsevier; 2011. pp. 179-218

[196] Reidy B, Haase A, Luch A, Dawson K, Lynch I. Mechanisms of silver nanoparticle release, transformation and toxicity: A critical review of current knowledge and recommendations for future studies and applications. Materials. 2013;**6**(6):2295-2350

[197] Sharma VK. Stability and toxicity of silver nanoparticles in aquatic environment: A review. In: Shamim N, Sharma VK, editors. Sustainable Nanotechnology and the Environment: Advances and Achievements. Washington, DC: American Chemical Society; 2013. pp. 165-179

[198] Yu KP, Huang YT, Yang SC. The antifungal efficacy of nano-metals supported TiO_2 and ozone on the resistant *Aspergillus niger* spore.

Journal of Hazardous Materials. 2013;**261**:155-162

[199] Sütterlin S, Tano E, Bergsten A, Tallberg A, Melhus H. Effects of silver-based wound dressings on the bacterial flora in chronic leg ulcers and its susceptibility in vitro to silver. Acta Dermato-Venereologica. 2012;**92**(1):34-39

[200] Dragieva I, Stoeva S, Stoimenov P, Pavlikianov E, Klabunde K. Complex formation in solutions for chemical synthesis of nanoscaled particles prepared by borohydride reduction process. Nanostructured Materials. 1999;**12**(1-4):267-270

[201] Hamouda T, Myc A, Donovan B, Shih AY, Reuter JD, Baker JR. A novel surfactant nanoemulsion with a unique non-irritant topical antimicrobial activity against bacteria, enveloped viruses and fungi. Microbiological Research. 2001;**156**(1):1-7

[202] Rahban M, Divsalar A, Saboury AA, Golestani A. Nanotoxicity and spectroscopy studies of silver nanoparticle: Calf thymus DNA and K562 as targets. The Journal of Physical Chemistry C. 2010;**114**(13):5798-5803

[203] Hackenberg S, Scherzed A, Kessler M, Hummel S, Technau A, Froelich K, et al. Silver nanoparticles: Evaluation of DNA damage, toxicity and functional impairment in human mesenchymal stem cells. Toxicology Letters. 2011;**201**(1):27-33

[204] Lim HK, Asharani PV, Hande MP. Enhanced genotoxicity of silver nanoparticles in DNA repair deficient mammalian cells. Frontiers in Genetics. 2012;**3**

[205] Ellinger-Ziegelbauer H, Aubrecht J, Kleinjans JC, Ahr H-J. Application of toxicogenomics to study mechanisms of genotoxicity and carcinogenicity. Toxicology Letters. 2009;**186**(1):36-44

[206] Dakal TC, Kumar A, Majumdar RS, Yadav V. Mechanistic basis of antimicrobial actions of silver nanoparticles. Frontiers in Microbiology. 2016;**7**:1831

[207] Li WT, Chang HW, Yang WC, Lo C, Wang LY, Pang VF, et al. Immunotoxicity of silver nanoparticles (AgNPs) on the leukocytes of common bottlenose dolphins (*Tursiops truncatus*). Scientific Reports. 2018;**8**(1):5593

[208] Pellieux C, Dewilde A, Pierlot C, Aubry J-M. Bactericidal and virucidal activities of singlet oxygen generated by thermolysis of naphthalene endoperoxides. Methods in Enzymology. 2000;**319**:197-207

[209] Kim JY, Sungeun K, Kim J, Jongchan L, Yoon J. The biocidal activity of nano-sized silver particles comparing with silver ion. Journal of Korean Society of Environmental Engineers. 2005;**27**:771-776

[210] Kim JS, Kuk E, Yu KN, Kim JH, Park SJ, Lee HJ, et al. Antimicrobial effects of silver nanoparticles. Nanomedicine: Nanotechnology, Biology and Medicine. 2007;**3**:95-101

[211] Kim S, Choi JE, Choi J, Chung K-H, Park K, Yi J, et al. Oxidative stress-dependent toxicity of silver nanoparticles in human hepatoma cells. Toxicology In Vitro. 2009;**23**(6):1076-1084

[212] Kim SH, Lee HS, Ryu DS, Choi SJ, Lee DS. Antibacterial activity of silver-nanoparticles against *Staphylococcus aureus* and *Escherichia coli*. Korean Journal of Microbiology and Biotechnology. 2011;**39**:77-85

[213] Wu D, Fan W, Kishen A, Gutmann JL, Fan B. Evaluation of the antibacterial efficacy of silver nanoparticles against *Enterococcus faecalis* biofilm. Journal of Endodontics. 2014;**40**(2):285-290

[214] Vallyathan V, Shi X. The role of oxygen free radicals in occupational and environmental lung diseases. Environmental Health Perspectives. 1997;**105**:165

[215] Thannickal VJ, Fanburg BL. Reactive oxygen species in cell signaling. American Journal of Physiology. 2000;**279**:L1005-L1028

[216] Stambe C. The role of p38 mitogen-activated protein kinase activation in renal fibrosis. Journal of the American Society of Nephrology. 2004;**15**(2):370-379

[217] Shrivastava S, Bera T, Roy A, Singh G, Ramachandrarao P, Dash D. Characterization of enhanced antibacterial effects of novel silver nanoparticles. Nanotechnology. 2007;**18**(22):225103

[218] Yang W, Shen C, Ji Q, An H, Wang J, Liu Q, et al. Food storage material silver nanoparticles interfere with DNA replication fidelity and bind with DNA. Nanotechnology. 2009;**20**(8):085102

[219] Marambio-Jones C, Hoek EMV. A review of the antibacterial effects of silver nanomaterials and potential implications for human health and the environment. Journal of Nanoparticle Research. 2010;**12**(5):1531-1551

[220] Le A-T, Le TT, Nguyen VQ, Tran HH, Dang DA, Tran QH, et al. Powerful colloidal silver nanoparticles for the prevention of gastrointestinal bacterial infections. Advances in Natural Sciences: Nanoscience and Nanotechnology. 2012;**3**(4):045007

[221] Soo-Hwan K, Lee HS, Ryu DS, Choi SJ, Lee DS. Antibacterial activity of silver-nanoparticles against *Staphylococcus aureus* and *Escherichia coli*. Korean Journal of Microbiology and Biotechnology. 2011;**39**:77-85

[222] Cheng X, Zhang W, Ji Y, Meng J, Guo H, Liu J, et al. Revealing silver cytotoxicity using Au nanorods/Ag shell nanostructures: Disrupting cell membrane and causing apoptosis through oxidative damage. RSC Advances. 2013;**3**(7):2296

[223] Asharani PV, Mun GLK, Hande MP, Valiyaveettil S. Cytotoxicity and genotoxicity of silver nanoparticles in human cells. ACS Nano. 2009;**3**(2):279-290

[224] Huang C-C, Aronstam RS, Chen D-R, Huang Y-W. Oxidative stress, calcium homeostasis, and altered gene expression in human lung epithelial cells exposed to ZnO nanoparticles. Toxicology In Vitro. 2010;**24**(1):45-55

[225] Xie H, Mason MM, Wise JP. Genotoxicity of metal nanoparticles. Reviews on Environmental Health. 2011;**26**(4):251-268

[226] Pilger A, Rüdiger HW. 8-Hydroxy-2′-deoxyguanosine as a marker of oxidative DNA damage related to occupational and environmental exposures. International Archives of Occupational and Environmental Health. 2006;**80**(1):1-15

[227] Valavanidis A, Vlachogianni T, Fiotakis C. 8-Hydroxy-2′-deoxyguanosine (8-OHdG): A critical biomarker of oxidative stress and carcinogenesis. Journal of Environmental Science and Health, Part C. 2009;**27**(2):120-139

[228] Ahmadi F. Impact of different levels of silver nanoparticles (Ag-NPs) on performance, oxidative enzymes, and blood parameters in broiler chicks. Pakistan Veterinary Journal. 2012;**26**:325-328

[229] Yan X, He B, Liu L, Qu G, Shi J, Hu L, et al. Antibacterial mechanism of silver nanoparticles in *Pseudomonas*

aeruginosa: Proteomics approach. Metallomics. 2018;**10**:557

[230] Alt V, Bechert T, Steinrücke P, Wagener M, Seidel P, Dingeldein E, et al. An in vitro assessment of the antibacterial properties and cytotoxicity of nanoparticulate silver bone cement. Biomaterials. 2004;**25**(18):4383-4391

Chapter 8

The Potential Application of Nanoparticles on Grains during Storage: Part 2 – An Overview of Inhibition against Fungi and Mycotoxin Biosynthesis

Daniel Nsengumuremyi, Parise Adadi, Gavers K. Oppong, Nadezhda V. Barakova and Elena F. Krivoshapkina

Abstract

Mycotoxins are secondary metabolites synthesized by filamentous fungi. They are common food contaminants that cause mycotoxicosis in humans and animals. Due to the severity of health risk pose by these mycotoxins, many countries have enacted strict measures to curb this menace. One promising measure is the use of nanoparticles. Herein, we present an overview of the application of titanium dioxide, chitosan, ultradisperse humic sapropel suspension, and carbon-based nanoparticles, a novel and innovative method of reducing mycotoxin production and the subsequent contamination of grains. All nanoparticles considered enhanced cell permeability by disrupting the membrane, resulting in the outflow of cellular materials. However, concentration, volume, type, and illumination (sunlight) influenced the fungicidal potential of NPs.

Keywords: filamentous fungi, mycotoxins, nanoparticles, fungicide, reactive oxygen species

1. Introduction

Microorganisms, including fungi, contaminate grains during storage. These fungi do not only reduce grain quality, but also produce mycotoxins which pose health risks to consumers [1, 2]. According to Kady et al. [3], *Aspergillus, Fusarium, Penicillium,* and *Rhizopus* are the most common genera in barley, wheat, maize, and sorghum. These grains serve as staple food worldwide. Nowadays, nanotechnology is advancing in many fields, namely biotechnology, analytical chemistry, agriculture, and others. However, its application in crop protection is still in its early stages [4, 5].

The biocidal activity of nanoparticles is well documented. Herein, we proposed the utilization of nanoparticles to inhibit fungal growth and the production/synthesis of mycotoxins. Therefore, the second part of this chapter aims to discuss other

IntechOpen

promising nanoparticles (titanium dioxide nanoparticles, chitosan nanoparticles, ultradisperse humic sapropel suspension (UDHSS) nanoparticles, and carbon-based nanoparticles/nanomaterials) of interest which could be applied during grain storage. The toxicological aspects, as well as the proposed modes of application are discussed.

2. Titanium dioxide nanoparticles

Titanium dioxide (TiO_2) nanoparticles (TiO_2-NPs), or ultrafine TiO_2, are particles of TiO_2 with diameters 1–100 nm. The TiO_2-NPs activity is exciting to researchers because of its specific characteristics which include; size, shape, crystal structure, surface stability among others [6]. They are among top five NPs used in consumer items such as cosmetics, food products, paints, and medicines [7]. TiO_2 received USFDA approval hence regarded as safe. It is widely used as food colorant in candies, sweets, chewing gums, etc. Anatase (used in printing inks and photocatalysts), rutile (used in colorants and sunscreens), and brookite are the three primary forms of TiO_2-NPs [8–12]. In 1985, Matsunaga et al. [13] first documented the antimicrobial activity of TiO_2. They observed that microbial cells were dead when exposed to a TiO_2-Pt catalyst illuminated with UV light.

The biocidal activity of TiO_2 has been reported [14–19]. **Table 1** shows the fungicidal activity of TiO_2-NPs against fungi species known to contaminate grains with the mycotoxins they synthesize.

TiO_2-NPs have been widely applied as antimicrobial agents in recent years due to their unique properties such as resistance to high temperatures, low solubility, high surface area, cost-effectiveness, hydrophilicity, and strong oxidizing properties [20].

TiO_2-incorporated polyethylene (PE) film inhibited growth of *E. coli* and *S. aureus*. UV light significantly enhanced the biocidal activity within 60 minutes of illumination [20]. Several studies [21–26] have documented the biocidal efficacy of TiO_2 against *E. coli*, *S. aureus*, *P. aeruginosa,* and *P. expansum*.

The photocatalytic oxidation of surfaces coated with TiO_2 and ultraviolet A (UVA) was effective against *E. coli*, *P. aeruginosa*, *S. aureus*, and *E. faecium* than the control [27]. A collaborated research [28] assessed the biocidal activity of the crude and annealed TiO_2-NPs. The results revealed that doped Ag-TiO_2 (7%) NPs killed 100%, 95%, and 96% of *P. aeruginosa*, *S. aureus*, and *E. coli,* respectively, at 40 mg/30 mL.

Assessing ecotoxicity of TiO_2-NPs against bioluminescent bacterium (*Aliivibrio fischeri*), algae (*Pseudokirchneriella subcapitata*, *Scenedesmus subspicatus,* and *Chlorella vulgaris*), protozoon (*Tetrahymena pyriformis*), water flea (*Daphnia magna*), and an aquatic macrophyte, *Lemna minor* [29] revealed these organisms showed significant behavioral and physiological changes when exposed to low TiO_2-NP concentrations (0.1 and 0.05 μg/L), thus demonstrated the ability of TiO_2-NPs to alter molecular pathways via which these organisms obtained vital nutrition for growth and synthesis of compounds (i.e., chlorophyll, etc.).

Maneerat and Hayata [26] tested the fungicidal activity of TiO_2 photocatalysts against *P. expansum* in the form of TiO_2 powder and TiO_2 coated on a plastic film. Both TiO_2-NPs suppressed the conidial germination and growth of the fungi. The quantity of TiO_2-NPs added correlated with the fungicidal activity.

Nitrogen-doped TiO_2 [TiO_2 (N)] exhibited potent biocidal activity with regards to reducing the number of surviving organisms than carbon-doped TiO_2 [TiO_2 (C)]. Therefore, TiO_2 (N) NPs can inactivate spores of *B. anthracis* (hazardous

Organism	Reference
C. albicans, *S. cerevisiae*	[31]
A. niger AS3315	[32]
F. verticillioides	[33]
A. niger spores	[34]
A. niger, *S. cerevisiae*	[35]
F. oxysporum f. sp. *lycopersici*	[36]
C. albicans ATCC 10231, *F. solani* ATCC 36031	[37]
C. albicans	[27]
C. famata	[38]
C. vini, *Hansenula anomala* CCY-138-30	[39]
Cladobotryum varium, *Trichoderma harzianum*, *Spicellum roseum*	[40]
Cladosporium cladospoiroides, *Epicoccum nigrum*, *F. mucor*, *Penicillium oxalicum*, *Trichoderma asperellum*, *Pestaotiopsis maculans*	[41]
Diaporthe actinidae	[25]
Erysiphe cichoracearum, *Peronophythora litchii*	[42]
Molds and yeasts (not specified)	[43]
Fusarium spp. (*equisetii*, *oxypartan*, *anthophilum*, *verticillioides*, *solani*)	[44, 45]
P. citrinum	[46, 47]
P. expansum	[26]
S. cerevisiae	[13, 48]

Modified with permission from Ref 4498160008350.

Table 1.
Fungicidal activities of TiO_2-NPs on mycotoxins-producing fungi

microorganism) under illumination by conventional light sources such as incandescent lamps [30].

2.1 Mechanistic action of TiO_2-NPs antimicrobial activity

TiO_2-NPs are the photocatalysts used to destroy unwanted organic compounds in the air, water, soil, and, more recently, in food [21].

Photocatalysis can be defined as the catalyst-driven acceleration of a light-induced reaction [49–52]. Homogeneous and heterogeneous photocatalytic processes utilize metal complexes (transition metal complexes like iron, copper, chromium, etc.) and semiconducting materials such as TiO_2, ZnO, SnO_2, and CeO_2 as catalysts. In the presence of light and heat, metal complexes become excited and form metal ion complexes, in contrast, semiconducting materials become excited due to the combination of electronic structures which is characterized by a filled valence band, empty conduction band, and light absorption properties, resulting in the generation of reactive oxygen species (ROS) or hydroxyl radicals. These hydroxyl radicals inflict damage to microbial cells [49–51, 53–55]. The subsequent hole in the valence band could further react with H_2O in the grains or hydroxide ions adsorbed on the surface of TiO_2-NPs to generate hydroxyl radicals (OH•), with electron in the conduction band reduce O_2 to superoxide ions (O_2^-) [21]. Gogniat and Dukan [56] demonstrated that DNA was denatured by hydroxyl radicals generated via the Fenton reaction resulting in cell death.

Electron paramagnetic resonance (EPR) spectroscopy study confirmed the photoproduction of hydroxyl radicals (OH•) from different TiO_2. The efficiency of hydroxyl radical generation depends on the source/origin of TiO_2 [57].

Cells are negatively charged [58] under optimum physiological condition due to heparan sulfate proteoglycans [59]. However, disease could trigger the cells to synthesize certain compounds which cause cell surface to become positively charge. Microbial cell could act as a hole for electron transfer between organism and its components [60]. The iron cluster on cell surface, in the periplasmic space, or inside the cell (proteins (such as ferritin)), could act as a precursor for iron-catalyzed Haber-Weiss reaction, which generates additional hydroxyl radicals in the presence of H_2O_2 and the superoxide ion [61].

Different treatments (photocatalysis, water, TIO_2, UV-A) applied to elucidate the effects of lipid peroxidation on *S. cerevisiae* revealed high malondialdehyde (MDA) in TiO2 -treated subjects with 2 hours. The results demostrated that TiO_2 was sufficient to damage membrane, thus interfered with permeability of the cell which led to the leakage of vital intracellular molecules (**Figure 1**) [48]. Similarly, Draper and Hadley [62] found photocatalysis-induced cell wall damage on S. cerevisiae [48]. This may decrease intracellular enzymatic activity as well as leaking of amino acids and NH_4^+, suggesting a drastic impact on proteins [63].

Cellular respiratory enzymes lost their activity after been exposed to irradiated TiO_2 (0.5 mg/mL), and the kinetics correlated with the losses of cell viability. Furthermore, when glucose was used instead of succinate as the electron donor, similar effects were observed. From this outcome, Li et al. [78] proposed that ROS generated from an irradiated TiO_2 surface, interacted with the polyunsaturated phospholipids in *E. coli*. Moreover, cell membrane structure was perforated due to lipid peroxidation creating a hole for more TiO_2-NPs to pass into interior of the cell, thus rendering respiratory proteinsinactive and subsequent cell death.

A progressive decrease in esterase activity was observed after exposing *S. cerevisiae* to irradiated TiO_2 [63]. Other researchers documented overexpression and inhibition (expressed at lower levels, including those encoding six cbb3-type cytochrome C oxidase subunits, an electron transfer flavoprotein, and

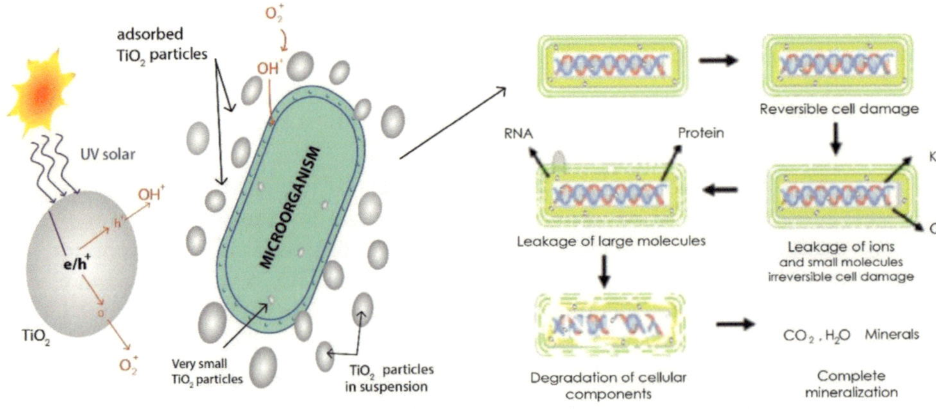

Figure 1.
Schematic illustration of the solar photocatalytic process for microbial cell inactivation in the presence of an aqueous suspension of TiO_2. Modified with permission from ref 4498160008350 [72]. Contact between the cells and TiO2-NPs affects membrane permeability; however, this is reversible. The availability of more NPs could enhance the damage to cell wall, thus allowing leakage of small molecules such as ions. Damage at this stage may be irreversible, and this accompanies cell death. Higher molecular weight components such as proteins could further be leaked followed by protrusion of the cytoplasmic membrane into the surrounding medium through degraded areas of the peptidoglycan and lysis of the cell. Intracellular components are then degraded progressively especially from the point of contact with photocatalyst, followed by complete mineralization.

two oxidoreductases) of genes associated with energy production and conversion processes. TiO_2-NPs exerted a stimulating effect on the respiratory chain and the electron transfer mechanism of the microorganism [64, 65].

Likewise, Matsunaga et al. [13] observed that incubating TiO_2/Pt NPs under metal halide lamp irradiation with *E. coli, Ch. vulgaris, L. acidophilus*, and *S. cerevisiae* inhibited cell respiration mechanisms and subsequent cell death. However, the results were not consistent as *Ch. vulgaris* had a thick cell wall mainly composed of polysaccharides and pectin hence, had comparative advantages (protection) over the other microbes.

Kubacka et al. [65] examined genome/proteome-wide expression profiles of *P. aeruginosa* PAO1 cells treated with TiO_2-based nanocomposite films. An increase and decrease in the levels of 165 and 151 transcripts were respectively reported in cells with TiO_2-coated Ethylene vinyl alcohol (EVOH) particles. Few proteins were detected at a statistically significant level ($p \leq 0.1$) in cells treated with TiO_2-coated EVOH particles compared to the control. TiO_2-UV treatment significantly suppressed (from 5.4- to 15.1-fold) the expression levels of genes essential for cell wall. However, 14 genes encoding for lipid metabolism essential for cell membrane were over-expressed (from 5.6- to 23.0-fold), unexpectedly, 2 were expressed at a lower level (from 5.5- to 7.4-fold).

In vivo and in vitro studies confirm that hydroxyl radicals inflict damage (breakage) on DNA strands. The extent of damage was minimized when dimethyl sulfoxide, catalase, or mannitol were incorporated in the reaction mixture [66]. However, the findings [66] contradicts previous studies [21, 67]. Exposing either purine or pyrimidine bases to TiO_2 and light from a 100-W Hg lamp resulted in the detection of NO_3^- and NH_4^+ ion. However, when native DNA and RNA molecules were subject to the same conditions, unknown peroxide species, along with phosphate and carbon dioxide, were detected, suggesting the breakage and mineralization of sugar-phosphate backbone of DNA and RNA molecules, respectively [68].

Kikuchi et al. [67] demonstrated the role of ROS on photocatalytic bactericidal activity. They utilized a porous polytetrafluoroethylene (PTFE) membrane in their system to physically separate the *E. coli* suspension from the TiO_2 thin. The results showed an impressive photokilling capability of the system with and without (control) PTFE - which was attributed to the generated H_2O_2. A group [69] demonstrated the stimulating effect of TiO_2-NPs on lipolytic activity in *A. niger*. The results showed that TiO_2-NPs significantly increased lipase biosynthesis (more than 1.5 times) compared to the control experiment. Treatment with TiO_2-NPs (size: 40 nm, concentration: 10 mg/L) in all culture media, enhanced lipolytic activity by 78.57% and 57.49% on the 4th and 5th day of cultivation, respectively. This finding reaffirms that smaller NPs can penetrate the cell membrane easily than bigger NPs, thus easily interact with molecular proteins, resulting in stimulating effects.

Gomes et al. [70] assessed the effects TiO_2-NMs (NM103, NM104, and NM105) and bulk TiO_2 against *Enchytraeus crypticus* with and without UV radiation. Microarray analysis revealed 10431 differentially expressed genes (DEGs) ($p < 0.01$) triggered as a result of exposure to TiO_2-NMs under no-UV. All samples under UV exposure registered an up-regulation of several transcripts, including caspase apoptosis-related cysteine peptidases, a signature of apoptosis activation, whereas under darkness the apoptotic signaling pathway was inhibited, suggesting that the oxi-radicals generated during the photoactivation of TiO_2 might substantially contribute to the apoptotic response and damage to the cell membrane. DNA damage was triggered after exposing samples to bulk/nano TiO_2 [71]. However, the findings of Gomes et al. [70] contradicted the [71] as reported that TiO_2-NMs_under no-UV impaired DNA repair, while bulk_TiO_2 under no-UV activated DNA repair mechanisms, suggesting that size of the TiO_2-NPs contributes to biocidal activity.

3. Chitosan nanoparticles

Chitin and chitosan have been widely used in the fabrication of polymer scaffolds [73]. Chitosan is a linear polysaccharide, a nontoxic biopolymer derived from the deacetylation of chitin, and used in many fields, including agriculture, medicine, and in vinification due to its biocidal potential. In agriculture, chitosan is used as biopesticide; in medicine, it is used to stop bleeding, wound healing, and as an antibacterial agent. Biodegradability, high permeability, nontoxic to humans, and cost-effectiveness are the features which make chitosan NPs unique. Chitosan and its derivatives have attracted considerable attention due to their biocidal activities [74, 75]. Several authors have reported the beneficial application of chitosan and its oligosaccharides which includes antitumor [76], neuroprotective [77], antimicrobial [78–85], and anti-inflammatory [86] agents. **Table 2** summarizes the fungicidal activities of chitosan against important agricultural microorganisms contaminating stored grains.

Fungal decay on pear fruit was suppressed by the combination of chitosan, yeast antagonist *Cryptococcus laurentii*, and $CaCl_2$. The results showed that mixture of chitosan at 0.5% and *C. laurentii* exerted greater effects compared to chitosan or *C. laurentii* alone. $CaCl_2$ showed little antifungicidal activity; however, it combination with chitosan and *C. laurentii* led to an effective and stable reduction of fungal decay [87], thus minimize or eradicate the menace of postharvest losses. Anthracnose in papaya caused by *Colletotrichum gloeosporioides* was controlled by the combination of *Burkholderia cepacia*, chitosan (0.75%) and $CaCl_2$ [88]. Postharvest blue, green, and grey molds affecting apple, oranges, and lemons were effectively controlled by mixing glycol chitosan (0.2%) with *Candida saitoana* [89–91]. Ag/chitosan-NPs showed significant antifungal activity against *A. flavus*, *A. alternata*, and *R. solani* hence could be used during grain storage [92, 93]. The synergistic effect (fungicidal activities) of hybrid copper(II) chitosan NPs to inhibit the growth of *F. graminearum*, *Verticillium dahlia* 57, and *F. solani* 169 was reported. In both cases, the NPs exerted an excellent efficacy in repressing the growth of fungi [94, 95]. Other authors reported that certain strains of *A. flavus*, *Cladosporium cladosporioides*, *P. aurantiogriseum*, and *Torulaspora delbrueckii* were resistant to chitosan at levels as high as 1% [7, 96]. The application of chitosan (0.025 and 0.05%) was effective against *Saccharomycodes ludwigii* and *Saccharomyces exiguous*. A rapid reduction in the number of yeast colonies was observed 2–4 min after application [97].

According to an earlier report, the effectiveness of the biocidal activity of chitosan depends on the molecular weight, degree of acetylation, and concentration [98, 99]. The application of NPs coated with polyethylene glycol (PEG) and natural garlic oil against *Tribolium castaneum*, a vital storage pest showed high efficiency over an extended period (8 months) due to the slow and persistent release of the active components [100]. The study highlighted the potential application of PEG-NPs as capsules to encapsulate various natural bioactive ingredients (i.e., oil from *Azadirachta indica*, extracts of *Khaya anthotheca*, alkaloid extracts of *Piper guineense* [101], etc.) for controll release and subsequent killing of microorganisms and pests during grain storage. Furthermore, [102, 103] extensively reviewed the literature on the biocidal activities of natural compounds (i.e., herbs, species, etc.) and its potential application in postharvest control.

3.1 Mechanistic action of chitosan nanoparticle antimicrobial activity

According to literature [116, 117], chitosan is composed of polycationic copolymers, with glucosamine and N-acetylglucosamine as axillary units, which contributes to its antimicrobial activity. The difference in environmental pH, pKa

DOI: http://dx.doi.org/10.5772/intechopen.93213

Reference	Sources of chitosan (CTS)	Deacetylation (%)	Microorganisms	Concentration	Form applied
[104]	Not reported (industrially made)	71.5	*A. niger*, *A. parasiticus*	0 (control), 0.1, 0.5, 1.0, 2.0, 3.0, 4.0, and 5.0 mg/mL	Solutions
[105]	Not reported (industrially made)	75–85	*A. flavus* IMI242687, *C. cladosporioides* IMI 274019, *M. racemosus IMI 017313*, *P. aurantiogriseum* IMI 297953, *Byssochlamys* spp. BF, *Byssochlamys* spp. GCB, *Byssochlamys* spp. SB, *S. cerevisiae* 28, *S. cerevisiae* 3085, *S. cerevisiae* SD, *Z. bailii*906, *Z. bailii* HP, *S. exiguus* 391, *S. pombe*, *S. ludwigii*	0, 1, 5, 10 g/L (fungi) and 5 mL (for yeast)	Solutions
[106]	Not reported	85, 81, and 82 for low-, medium-, and high-molecular weight chitosan respectively	*A. alternata*, *B. fabae*, *F. oxysporum*, *P. digitatum*, *P. debrianum*, *R. solani*	250, 500, 1000, 1500, 2000, 2500, 3000, 3500 and 4000 mg/L	Solutions
[97]	Not reported	79	*S. exiguus*, *S. ludwigii*, *T. delbrueckii*	0.05%, 0.005%	Solutions
[107]	Not reported	Not reported	*C. neoformans* strain *B3501*	Different concentration (0, 0.625, 1.25, 2.5, and 5 mg/mL) was employed	Biofilm
[108]	Industrially prepared chitosan	95	Psychrophilic, mesophilic, *Pseudomonad*, yeasts and molds	Not reported	Coating
[87]	Crab shell	~90	*P. expansum* (blue mold)	Various concentrations were applied for in vivo (0, 0.1, 0.5 and 1.0% (w/v); in vitro (0, 0.001, 0.01, 0.1, 0.2, 0.3, 0.4, 0.5, 0.6, 0.7, 0.8, 0.9 and 1.0% (w/v)) experiment	Coating
[109]	Crab shell	82	*F. oxysporum f.* sp. ***radicis-lycopersici*** 127, *Candida krusei* VKPM Y-2594, *E. coli* ATCC 5945, *P. aureofaciens*VKPMB-7542, *E. agglomerans* VKPM B-7541, *B. subtilis* VKPM B-7540	0.1, 0.5, or 1.0 mg/mL for the fungicidal test while 0.9 mL of LMW or DDC-LMW chitosan was used for the bacteriostatic test	Solutions

Reference	Sources of chitosan (CTS)	Deacetylation (%)	Microorganisms	Concentration	Form applied
[110]	Not reported	75-85	*R. stolonifera*, *E. coli* DH5α strain	1.0 mL	Coatings and solutions
[111]	Not reported (industrially made)	100 (MMW) and 97 (LMW)	*R. oryzae* CECT 2340, *A. alternata* CECT 20560, *A. niger* CECT 2088	6 mL	Films and solutions
[112]	Not reported (industrial made)	85–89	Molds and total flora isolated from strawberries (*R. Stolonifer* and *B. cinerea*)	Final concentration before the spraying was 0.02%, w/v	
[113]	Not reported	80%	*R. solani* Kuhn, *F. oxysporum* (Schl.) *f.* sp. *Cucumernum owen, C. cucumerinum Ell.* Et Arthur, *B. cinerea Pers.*, *C. orbiculare* (Berk. & Mont.) *Arx*, *P. asparagi* (sacc.a) *Bubak*, *A. Kikuchiama Tanaka*, *P. italicum Wehmer*, *Fusarium oxysporum Schl. F. Sp. Uasinfectum* (Atk.) Snyd. & Hans, *V. ctahliae Kleb.*, *R. sclani Kuhn.*, *B. berengeriana de Not. f.* Sp., *Piricola* (Nose) *Koganezaea et Sakuma*, *Sclerotinia sclerotiorum (Lib.) de Bary*, *Venturia nashicola Tanaka et Yamamoto*, *Gibberella zeae (Schw.) Petch* and *Phytophthora infestabs* (Mont.)	20, 30, 50, 100, and 150 mg/L	Solutions
[114]	Not reported	90	*A. niger*	0.1% or 1% (w/v)	Coatings, films, and liquid
[115]	Shrimp shell	Not reported	*A. alternata f.* sp. *lycopersici*	100–6400 μg/mL	Solutions

Table 2.
Some selected studies on fungicidal activities of chitosan NPs.

of chitosan and its derivatives creates an electric field for an electrostatic interaction between the polycationic structure and the anionic components of the cell (i.e., lipopolysaccharide and cell surface proteins), thus altering cell permeability [118–123]. High pH enhance rapid protonation, which increase the positive charge density (polycationic activity) of chitosan. A positive correlation was established between charge density and the biocidal activity of quaternized chitosan [124–127]. The inhibition potential of chitosan could be incapacitated when the charge density is reduced [120] due to changes of pH values. A similar outcome was reported by Qin et al. [128]. The antimicrobial mechanism was associated with the interaction of the negatively charged cell membranes and the cationic NH^{3+}groups of the chitosan derivative, which increase membrane permeability resulting in lysis [129] and leakage of macromolecules killing the cells. A carboxyfluorescein (CF)-loaded liposome study showed the effectiveness of lower molecular weight (LMW) chitosan on the cell membrane. The results showed that 0.75 μg/μL of LMW chitosan triggered moderate (≈7%) leakage of carboxyfluorescein found in the large unilamellar vesicles [130]. Similarly, Ing et al. [131] reported that chitosan NPs prepared from different concentrations of LMW and high molecular weight (HMW) showed efficient inhibitory activity against *C. albicans* (MIC_{LMW} = 0.25–0.86 mg/mL and MIC_{HMW} = 0.6–1.0 mg/mL) and *F. solani* (MIC_{LMW}= 0.86–1.2 mg/mL and MIC_{HMW} = 0.5–1.2 mg/mL) compared to the solution form (MIC = 3 mg/mL for both MWs and species). The authors established a statistical linear relationship between MW and particle size/zeta potential, thus provided an avenue for the manipulation of physicochemical properties of NPs to maximize its ability to penetrate the cells, trigger leakage of intracellular component, eventually killing the fungi and extend safety of the grains.

Researchers [132–135] proposed the fundamental mechanism contributing to interaction of negatively charged surface components of fungi and bacteria with the positively charged NH^{3+} groups of glucosamine (chitosan), which alters cell surface, and trigger leaking of intracellular substances, resulting in the impairment of vital physiological activities thus killing the microorganism. The inability of the second amino groups on N-acetylation of chitosan oligomers to donate positive chargeresult in the inhibition of its fungistatic activity [136]. Therefore, the contribution of NH^{3+} groups to biocidal activity cannot be ignored and should carefully be considered to maximize the effects.

The outer membrane (OM), inner core of lipopolysaccharide (LPS) molecules, and lipid components of Gram(−) bacteria are composed of anionic groups like phosphate and carboxyl, which contribute to the hydrophilic nature of the cell wall, thus creatin interaction of charges (electrostatic) with divalent cations. The OM protects Gram (−) bacteria cells from macromolecules and hydrophobic compounds (antibiotics and toxic drugs), giving Gram(−) bacteria a comparative advantage over Gram(+) bacteria. Therefore, breaching the integrity of the OM by chitosan could enhance its biocidal activity toward Gram(−) bacteria [137, 138]. On the other hand peptidoglycan (PG) and teichoic acid (TA) on the cell wall of Gram(+) bacteria have polyanionic group, which facilitates interaction via covalent bond with N-acetylmuramic acid in the PG layer, or via glycolipid- which links outer leaflet of the cytoplasmic membrane [139]. As documented by Kong et al. [120], the poly(glycerol phosphate) anion groups aid the structural stability of cell wall in addition to some membrane-bound enzymes.

LMW chitosan showed higher efficiency perforate/penetrate the microbial cell compared to HMW chitosan, which interacts with DNA to change the translation and transcription profile of genes. Chitosan binds to DNA with accurate precision, denying the organism of normal DNA transcription and mRNA synthesis, resulting in cell death [140–142].

A decrease in the induction of β-galactosidase was observed when yeast cells were exposed to chitosan. A concentration of 0.35 mg/mL chitosan reduced β-galactosidase activity by 32%. An increased in concentration (1.25 mg/mL) further led to the reduction of enzyme activity. The control experiment did not follow the trend. Likewise, the treated cells showed that chitosan greatly influenced protein biosynthesis in the yeast [130]. Previous work [143] documented cell sensitivity to chitosan, which altered the deletions of genes involved in sphingolipid (e.g., *ipt1Δ*, *skn1Δ*, *lcb3Δ*) and ergosterol (e.g., *erg3Δ*, *erg5Δ*) biosynthesis. In 1981, Hadwiger et al. [144] detected chitosan within plant cytoplasm and nucleus within 15 min after application, which indicate that chitosan can efficiently penetrate the thicker cell wall (the reason for its detection) and potentially interfered with DNA transcription and translation. This study suggests that chitosan can easily penetrate microbial cells since plants have a thicker cell wall than microbes.

Moreover, looking at the time factor (15 min), it is evident that chitosan can quickly interact with fungi and bacteria cellular DNA with subsequent inhibition of DNA transcription, as well as RNA and protein synthesis [140, 145, 146], leading to cell death. Chitosan triggered transcriptional responses when introduced to *S. cerevisiae* strain X2180-1A (MATa SUC2 mal gal2 CUP1). T-Profiler analysis showed cis-regulatory motifs apart from the environmental stress response correlated positively with expression in the chitosan-treated sample. Cin5p, Crz1p, and Rlm1p were the transcription factors associated with identified binding sites. Genes participating in cell wall organization, biogenesis, and signal transduction were also triggered in the treated sample compared to the control [134]. Some factors influencing the antimicrobial activity of chitosan is discussed above; however, Kong et al. [120] and Hosseinnejad and Jafari [147] published an excellent reviews on these factors.

4. Ultradisperse humic sapropel suspension (UDHSS) nanoparticles (UDHSS-NPs)

Sapropel is benthos found in fresh water, formed under anaerobic conditions from dead organic matter of anhydrobiotic microflora and microfauna. It is principally composed of nutrients (i.e., sugars, minerals, lipids, etc.) and organic compounds known as humic substances [148–151]. Sapropels and sapropel extracts showed antibacterial and antifungal properties in previous studies hence could used as an alternate and novel biocidal agent during grain storage. The antimicrobial properties of sapropels is attributed to the presence of humic substances [152–156]. Sapropel has become a popular raw material for therapeutic applications, production of sorbents, organic fertilizers, and food supplements [157]. UDHSS-NPs are organic NPs which exhibits potent biocidal activities due to the presence of humic substances [148]. Fulvic acids (FAs), humic acids (HAs), mumie, and humin are the principal constituents of humic substances (HSs) in sapropels [158–161], and are reportedly attribute to their biocidal properties. Many studies [152–155] have illustrated the inhibitory effects of sapropel on bacteria (*S. aureus*, *E. coli*, etc.) and yeasts (*Candida*, etc.). A micrograph of UDHSS-NPs is shown in **Figure 2** however, its characteristics were not included in the present study.

In a series of tests performed by Barakova et al. [148], experiments 2 and 3 exhibited most significant fungicidal effects on *A. niger*, a species which poses a greater threat to grain/food industries due to the potent mycotoxins it produce. A report showed that hematite NPs (hematite-HA complexes) significantly inhibited the growth and gene expression of *P. putida* KT2440. The bactericidal activities were ascribed to the oxidative stress induced by generated ROS. It was also shown that the physicochemical properties of the NPs (e.g., surface charge and size) influenced the efficacy of the hematite-HA complexes [162]. Therefore, modification of UDHSS-NPs could improve its biocidal properties.

A group of researchers [163] assessed the fungicidal activity of HAs and FAs extracted from soils on phytopathogenic fungal species (*Physalospora piricola* (P.P), *Botrytis cinerea* (B.C), *Rhizoctonia cerealis* (R.C), *Fusarium graminearum* (F.G), *Phytophthora infestans* (P.I), *Sclerotinia sclerotiorum* (S.S), *Rhizoctonia solani* (R.S), *Cercospora arachidicola Hori* (C.H), and *Bipolaris maydis* (B.M)). The results showed that HA exhibited above 30% and 50% inhibition against B.C, R.C, F.G, P.I, and P.P, respectively. The inhibition exerted by HA on all the species was higher compared to FA except for B.C. Correlation analysis further revealed that the inhibition rates of HAs decreased significantly with time (years) ($p < 0.05$) against most tested fungi except P.I., whereas FAs showed a negative correlation with cultivation years ($p < 0.05$) against most of the tested fungi except F.G. and S.S.

Recently, Ong et al. [164] documented that HAs (10 mg L^{-1} HA) altered enzyme activity in zebrafish embryo. Physicochemical properties such as size, zeta potential, and particle dissolution influenced their actions. It was further shown that coupling HAs with NPs enhanced the activity of the composite NPs. The addition of HAs reduced the hydrodynamic diameters of all examined NP suspensions except cadmium selenide (CdSe) NPs. Ezhkov and colleagues [165] developed NP-sapropel composite with particle size 45.0–180.0 nm and investigated its effects on treated albino mice. The results showed scarring of organ walls and shedding/exfoliation of the superficial epithelial cells. Further histological analysis of the oesophagus wall showed a significant thinning of the horny substance and the removal of the stratified epithelium of the mucous membranes in areas in contact with the NPs.

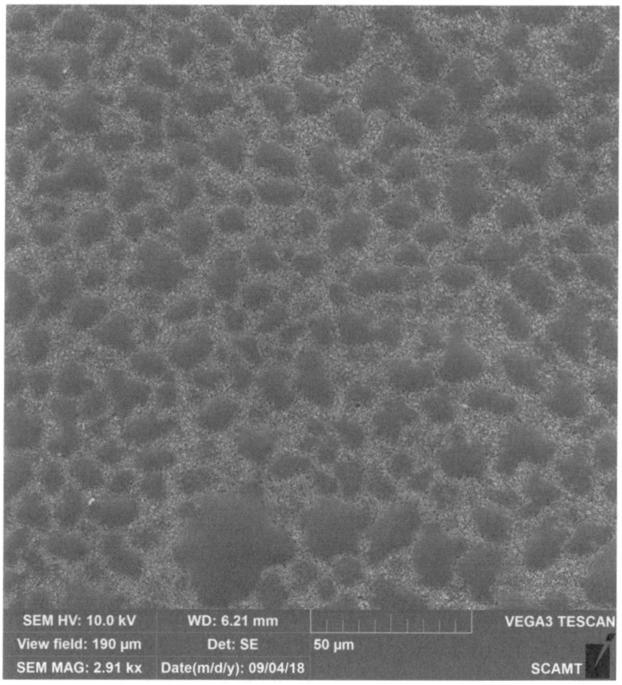

Figure 2.
UDHSS nanoparticles under a scanning electron microscope (SEM).

4.1 Mechanistic action of UDHSS-NPs antimicrobial activity

Several studies have described the biological activity of sapropel on enzymes, which confirms its antimicrobial activity. Details of these studies are discussed below in a quest to put forward a proposed mechanism by which UDHSS-NPs kill microorganisms. Environmental factors such as temperature, pH, oxygen, and moisture play a vital role in the mechanistic action of UDHSS-NPs. According to Perdue [166], HSs is complex mixture containing aliphatic, aromatic carboxyl and hydroxyl functional groups, which binds with microbial cells either on grains or in the environment (i.e., water, soil, etc.), thus alter the membrane structural intergrity and its functions. According to literature, the fungi cell walls share similarities with plant and bacterial and indeed with the extracellular matrix material of mammalian cells. The anionic surface, β1,4- and β1,3-linked polystarch forms a ribbon-like or helical (β1,3-glucan) structures which interacts with opposite charges. The cross linking of glycans of in eubacterial walls with peptides as well as phenolics and polysaccharides in plats promotes hydrogen bonding [167, 168]. Furthermore, the fungal cell wall is uniquely composed of mannoproteins, chitins, α- and β-linked glucans which serves many functions including; metabolism, ion exchange as well as providing cell rigidity and shape [169]. With the latter interacting with the HS. The interactions between HS and microbial cells depend on the lipophilicity and electric potential of the HS and cell [170], coupled with the size of the UDHSS-NPs. Microbial cells are composed of cations such as H^+, Na^+, K^+, Li^{2+}, Al^{3+}, Ca^{2+}, Cu^{2+}, or Pb^{2+} which interact with UDHSS-NPs thus penetrate the cell. As documented by Lofts et al. [171], cation-HS interactions exert control on the reactivity of cation, including its bioavailability for further reaction. Studies have shown the effects of binding metals with HS on water and soil ecosystems [172–174]. Natural and artificial HS got attracted to rice cells [175], macrophyte of *Ceratophyllum demersum*, crustaceans—*Gammarus pulex*, and vertebrates—tadpoles of *Rana arvalis* [176], which support the hypothesis that HS is charged and naturally interacts with microorganisms. When HS penetrates or is taken up by a cell, the electric potential of the cell is disrupted, denying the cell the ability to provide support in terms of rigidity, shape and metabolism, thus creating pores through which vital intracellular structures are leaked out.

In an in vivo experiment, Vigneault et al. [177] discovered that Suwannee River HA and FA enhanced the release/leakage of the fluorescent probe sulforhodamine-B (SRB) encapsulated within 1-palmitoyl-2-oleoyl-sn-glycerol-3-phosphatidylcholine (POPC) vesicles. With regards to HA, a pH from 7.6 to 5.7 enhanced its surfactant-like effect. In conclusion, HS can alter the permeability of microbial cell, to create pores via which intracellular components are leaked out, killing the microorganism. However, the concentration, functionalisation (acylation), and pH of HS could potentially influence the biocidal activity.

According to Almatov and Akhmerov [178], 0.2–0.8 mg/mL mumie activated mitochondrial respiration and inhibited cellular succinate-oxidase and NADH-oxidase activity (mitochondrion). Similarly, mumie triggered the outflow of Ca^{2+} [160].

Previous studies [179–181] reported that mumie induced a dose-dependent elevation of superoxide dismutase, catalase, and glutathione peroxidase in rats. These enzymes are involved in the generation of ROS in an HA-induced antimicrobial or biological effects, which killed microorganisms and other grain storage pest.

A small-molecular size humic (LMSH) extracted from the feces of *Nicodrilus* and *Allolobophora rosea* enhanced the uptake of nitrate by plant roots and the accumulation of anions in the leaves. Further molecular analysis showed that LMSH influenced gene transcription in roots and long-distance effects in shoots as observed for *Mha2* and the *ZmNrt2.1* gene, respectively [182], which indicate HS can interfere with protein synthesis in microbes. FA and HA extracted from a podzol stimulated respiration in rat liver mitochondria at concentrations between 40 and 360 mg/L. Depending on the duration of contact with mitochondria, uncoupled oxidative phosphorylation may occur subsequently affecting the growth of the microorganism [183].

A product of oxidative DNA damage, 8-hydroxy-2′-deoxyguanosine increased significantly after treatment with HA, indicating the ability of HA to inflict damage on DNA. The endonuclease activity of the viral RNA polymerase was inhibited when it came in contact with HA [184]. The concentrations (5, 10, and 15 mgL^{-1}) of HA and its organic extract significantly increased luciferase reporter gene activity in H4IIE. luc cells in a dose-dependent manner, which affected various molecular processes [185], thus killing the cell. The addition of HA (300 mg kg^{-1}) to soil stimulated the growth of bot laurel plants and rhizospheric bacteria and actinomycetes. However, high dose (3000 mg kg^{-1}), exerted an inhibitory effects [186]. The effects of HS on the hormone of *Caenorhabditis elegans* [170, 187], the sex ratio of *Xiphophorus helleri* [188], and the change in biochemical parameters of amphipod [189] were reported. These studies reiterate the potential biological effects of HS on microorganisms at the molecular level thus making them vulnerably for UDHSS-NPs.

5. Carbon-based nanoparticles/nanomaterials

Recently, carbon-based nanomaterials/particles (CNPs), which include nanotubes (i.e., double- or single-walled carbon nanotubes (DWCNT/SWCNTs)), fullerenes, and graphene oxide (GO) (**Figure 3**), have gained attention due to their potent biocidal activities. According to literature, the biocidal potency of these novel NPs is influenced by their physical/chemical properties, high adsorptive potentials, size, large surface area, and colloidal stability under wide range of pH. Increasing the NPs' surface area led to a decrease in size, with concomitant increase in adsorption and absorption (into fungi cell), which improved interaction [190–196] with subsequent inhibition of fungal growth.

The mycelia biomass and aflatoxin biosynthesis in *A. flavus* NRRL 3251 was negatively influenced at 10 μg mL^{-1} of fullerene C_{60} (fullerols $C_{60}(OH)_{24}$). The effects (growth arrest) was concentration-dependent. However, the antioxidative activity of the furrerols declined over time [197]. Hao and colleagues [198] investigated the fungicidal potentials of metal (copper oxide (CuO), ferric oxide (Fe_2O_3), and TiO_2NPs) and carbon-based NPs (multiwalled carbon nanotubes, fullerene, and reduced graphene oxide) against *Botrytis cinerea*. The results showed that all the six NPs exhibited biocidal activity with 50 mg/L of fullerene showing the strongest antifungal effects.

Reduced graphene oxide (rGO) nanosheets inhibited the mycelial growth of *A. niger*, *A. oryzae,* and *F. oxysporum* with half maximal inhibitory concentrations (IC_{50}) of 500, 500, and 250 μg/mL, respectively. The fungicidal activity as ascribed to the sharp edge of the rGO [199] which inflict injury on the cells, resulting in leaking of the cell components. Another hypothesis is that the organic functional groups on the fungi cell wall chemically interact with the ROS in rGO [200], which halts the uptake of nutrient and excretion of waste metabolites eventually killing the fungi.

Among the six carbon nanomaterials (SWCNTs, MWCNTs, GO, rGO, C_{60}, and activated carbon (AC)) assessed for their fungicidal activity against pathogenic fungi (i.e., *F. graminearum* and *F. poae*), SWCNTs (500 μg/mL) exhibited the most potent activity, followed by MWCNTs, GO, and rGO respectively. However, the other two CNPs (C_{60} and AC) showed minimal activity, probably due to insufficient contact with fungal spores [201]. Conclusively, increasing the concentration of CNPs (62.5 < 125 < 250 < 500 μg/mL) increased the fungicidal potency. In a similar study, Wang et al. [202] reported that modifying the surface of MWCNTs with –OH,

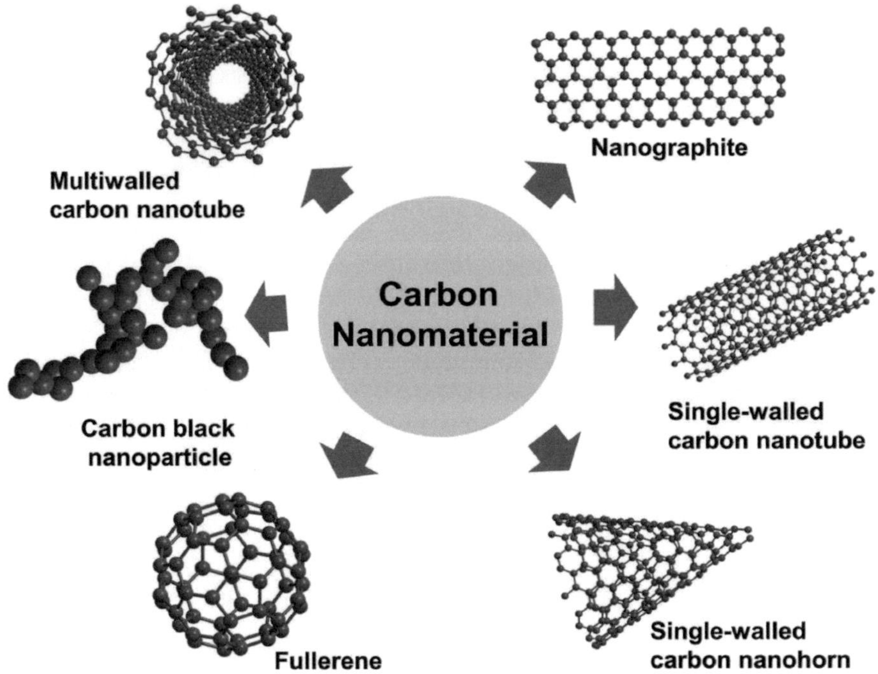

Figure 3.
Various carbon-based NPs [208].

–COOH, and $-NH_2$ improved the fungicidal activity (inhibition in spore elongation and germination) than the unmodified CNTs. It is hypothesized that modified CNTs formed a stable dispersions, which favoured interaction with spores, as a results enhanced antifungal activity. The authors observed a reduction in *F. graminearum* spore from 68.5, 54.5, 28.3, 27.4, and 29.5 μm, when 500 μg/mL MWCNTs (control), MWCNTs-COOH, MWCNTs-OH, and MWCNTs-NH_2 were applied, respectively. Moreover, previous works [203, 204] documented that biological activity of nanotubes improved upon addition of functionalized aliphatic amide (covalent) and polyethylene glycol (PEG) and/or polyoxyethylene(40)nonylphenyl ether (IGPAL) (non-covalent) chemical groups [205]. Zare-Zardini et al. [206] conducted a covalent functionalization of MWCNTs with lysine and arginine under radiation. The modified MWCNTs exhibited potent biocidal activity against all test fungi (A. *niger*, *A. fumigatus, C. albicans, P. chrysogenum, S. cerevisiae, F. culmorum, Microsporum canis, Trichophyton mentagrophytes, Trichophyton rubrum, and P. lilacinum*) compared unmodified MWCNTs. Surprisingly, the fungicidal activity of MWCNTs-arginine against all the test fungi was slightly higher than MWCNTs-lysine. The authors hypothesized that the positive charge on arginine might have enhanced the binding of NPs on the fungal membrane and altered the genetic makeup (DNA). Thus, lysine and arginine could be utilized to improve the fungicidal activity of CTNs. Recently, Katerine et al. [207] reported the fungicidal activity of cotton fabric silica-silver carbon-based hybrid NPs against *A*. sp., *Cladosporium* sp. and *Chaetomium globosum*. The fabrics with high number carbon exerted the most increased biocidal activity on *C. globosum* and *Aspergillus* sp.

5.1 Mechanistic action of CNPs fungicidal activity

The ability of CNPs to interact and integrate into fungi cells determines their fungicidal activities. Wang et al. [201] reported the importance of surface contact of CNPS to their biocidal functionality. A transmission electron microscopy study showed CNPs interacted and integrated into spores and form an aggregation. It was hypothesized that the van der Waals force in CNTs was strong enough formed a bond with spores, as a result, triggered plasmolysis. Similarly, Zare-Zardini et al. [206] reported a strong interaction in functionalized CNTs with arginine than lysine with fungi membrane. These interactions may result in CNPs been internalized into fungi. Transmission electron microscopy (TEM) analysis showed direct evidence of nanographene been internalized in Caco-2 cells [209], which support the above finding. Moreover, cells treated with CNPs showed evidence of plasmolysis [201].

They compared images of healthy and treated (incubation for 3 h with CNPs) cell membranes of *F. graminearum* spores. They observed an intact, slick, compact, inerratic, and well-positioned cytoplasm for the untreated cell; however, after treatment the latter cell were transfigured, contracted and gathered. This shows solute lost through CNPs point of contact. Interaction of CNPs with fungi' membrane led to a decrease in membrane integrity by the stresses exerted by the electrostatic forces between the microbial outer surface and CNTs, resulting in membrane oxidation [210]. CNPs are reported to be a contributing factor to the over-generation of ROS, which could trigger fungal cells to enter oxidative stress, causing excessive impairment to cellular components and permanent DNA laddering that could potentially lead biocidal activity against the cells [211, 212]. In contrast, Saha et al. [209] found that all CNPs (C1, C2, C3, C4 C5) assessed did not contribute to ROS production in Caco-2 cells. A decrease in ATP was observed. Conclusively, CNTs applied was internalized and disrupt the functionality of mitochondria which explains the reason for the low ATP observed. However, the treatment did not influence the production of ROS.

6. Toxicological aspects of NPs

According to Higashisaka et al. [213], NPs with diameters ≤100 nm are presently been used in various applications, including food production (e.g., to improve texture). An orally ingested NP can cross the gastrointestinal barrier, absorbed into the blood, and alter normal physiological functions, thus causing adverse health is to consumer(s) [213, 214]. Ezhkov and colleagues reported acute catarrhal inflammation on esophagus, stomach, and duodenum of mice fed with sapropel-NPs at a dose of 1.8 g/kg. However, 0.3 g/kg and 1.5 g/kg dose did not manifest any toxic effects [165].

A positive correlation was established between residues of Ag-NPs in rat organs and the NP suspension applied. NPs migrated from the luminal side to the intestinal epithelial cells via endocytosis or transcytosis, which are accumulated in the mentioned organs. However, all treated rats were able to excrete the NPs from most organs except the brain and testes [214–217]. Cellular uptake of NPs is similar to mechanism of the antimicrobial activity as its also depends on size, surface charge, and dispersion or aggregation state [218, 219]. Rhodamine B (RhB) labeled carboxymethyl chitosan grafted NPs (RhBCMCNP) and chitosan hydrochloride grafted NPs (RhB-CHNP) bearing positive or negative charges used as model chitosan to elucidate the effects of particle size and surface charge on the cellular uptake of NPs revealed that the surface charges were attracted to the macrophages, and could be attributed to the electrostatic interactions between particles and phagocytic cells. Besides, different cell lines, irrespective particle size, and surface charge difference influence the uptake of NPs [219]. Kim et al. [220] detected traces of Ag-NPs in blood, liver and other organs after they orally fed rats at a dose of 30, 125, or 500 mg/kg BW/day. A significant ($p < 0.05$)

dose-related decrease in the bodyweight of high-dose male rats at the fourth, fifth, and seventh weeks was observed; however, no significant dose-dependent changes in the female rats. Further hematological assays showed a significant increase ($p < 0.01$) in cholesterol in the both high-dose male and female rats. A significant increase ($p < 0.01$) in alkine phosphotase (ALP) was also indicated for the high-dose female rats. The authors reported no-observed-adverse-effect level (NOAEL) and lowest-observed-adverse-effect level (LOAEL) as 30 mg/kg BW/day and 125 mg/kg BW/day, respectively. Treatment with NPs of diameters ranging from 25 to 80 nm at a dose of 5000 mg/kg body weight altered the levels of alanine transaminase, aspartate transaminase, blood urea nitrogen, and lactate dehydrogenase, along with lesions on the liver and kidneys of female mice. Myocardial damage associated with groups showing a notable changes in serum LDH and alpha-HBDH levels compared to the control experiment. Also, a biodistribution test disclosed that TiO_2 was predominantly retained in the liver, spleen, kidneys, and lung tissues, indicating TiO_2-NPs were transported via endocytosis to other tissues and organs after their uptake by the gastrointestinal tract [221]. Contradictory finding was reported by Warheit et al. [222] where no adverse effects were manifested after orally fed rats with TiO_2-NPs. However, the NOAEL on rats exposed for 90 days was >1000 mg/kg BW/day. In a similar study, Sharma et al. [223] divided male Swiss albino mice into three groups (group 1—vehicle control (water); group 2—ZnO nanoparticles (300 mg/kg body weight); group 3—ZnO nanoparticles (50 mg/kg)) and fed them with 50 and 300 mg/kg b.wt. ZnO-NPs for 14 consecutive days. ZnO-NPs induced oxidative stress, which damage the DNA and apoptosis in the mouse liver. Additionally, elevated levels of ALT and ALP serum and subsequent pathological lesions were observed in the treated mice. Lastly, at a higher dose (300 mg/kg) of ZnO-NPs, a significant ($p < 0.05$) induction of lipid peroxidation was observed in the liver, brain, and kidney (**Figure 4**) of the treated mice in comparison with the control test. Cho et al. [224] discovered ZnO-NPs had a higher absorption efficiency than TiO_2-NPs in rats. ZnO-NP concentrations in the liver and kidney were significantly higher compared to the control, whereas with TiO_2-NPs, no dramatic increase was detected in the sampled organs. In the feces, very high and low concentrations of Ti and Zn were detected, respectively. The concentration of ZnO in the spleen and brain was minimally elevated. Similarly, Ti concentrations were not drastically increased in urine; in contrast, it was Zn levels, that remarkably

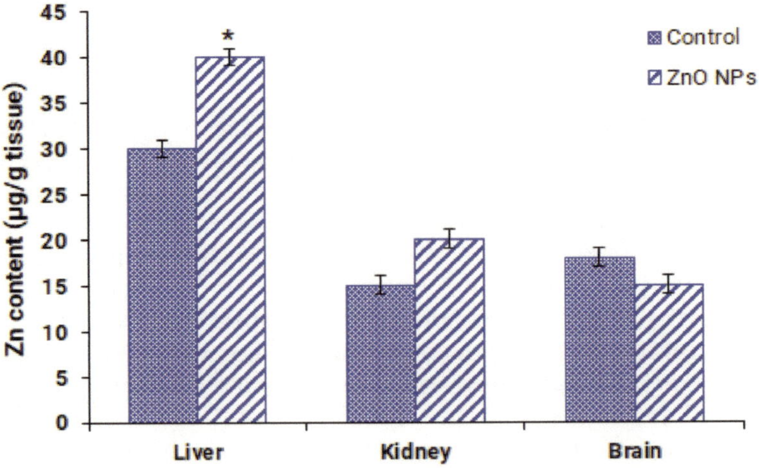

Figure 4.
Zinc content in selected tissue of the mice (n = 5) after oral administration of ZnO nanoparticles (NPs) (300 mg/kg) for 14 consecutive days. Data represent mean±S.E.M. of three animals. *$p < 0.05$*, compared to control. Modified with permission from ref 4495441125809.*

changed. Therefore, the absorption of various NPs could be attributed to the higher dissolution rate in the acidic gastric fluid; however, this might not be applicable when NPs are utilized during grain storage. When a stored grain undergoes sun drying, milling, etc., the levels of NPs may decrease to a level that could not affect the consumer health. Moreover, many NPs have received approval for application in many fields. Nevertheless, rigorous studies are warranted to expound on any risks or the safety of NPs use in grain storage. According to Zare-Zardini et al. [206], CNPs appeared less toxic to humans and animals compared to metal NPs and are therefore the better alternative and a novel method for reducing mycotoxin biosynthesis in grains.

7. Proposed methods of applying NPs during grain storage

1. The first method is direct processing of grains with solutions of the required concentration of NPs. To achieve this, biocompatible NPs in an aqueous dispersion medium with pH values close to neutral should be used.

2. Treatment with aerosols NPs could also be used if the NPs are dispersed evenly over the granary or silos. The aerosols to apply should be modified to prevent aggregation on grains. Using aerosols saves time and labor since additional drying is not required.

3. The use of packages made from NPs during storage, transportation, and sale will extend the shelf life of grains. Alternatively, NPs formulated cubes could be place in jute bags with grains; however, periodic mixing is required to distribute the NPs.

4. In our opinion, one of the most inexpensive methods is the use NPs in the production of materials for granaries, as well as treating interior and exterior surfaces of the storage facilities.

8. Conclusions

The ability of NPs to suppress the synthesis of mycotoxins in fungi and other microorganisms could be a breakthrough to curbe the issue of aflatoxin prevalence worldwide. NPs displayed excellent antifungal activity against important fungal species which contaminate grains with toxins during storage. The concentration, volume, type, and illumination (sunlight) significantly influenced the biostatic activity of NPs. Hence, these factors should carefully be considered when applying NPs in grain storage. The proposed NPs are environmentally friendly and pose no threat to consumer compared to some conventional methods of grain preservation. Several ex vivo, in vivo, and in vitro studies supports these claims. Moreover, NPs are biocompatible to the human system hence their usage in the food industry. Despite safety of NPs guaranteed by international safety organizations such as the Food Safety Authority, routine testing is required to understand the impact it has on grain nutritional, sensory, and other physicochemical parameters.

Acknowledgments

This work was supported by the Ministry of Science and Higher Education of Russia (Project No.075-15-2019-1896).

Disclosure statement

The authors reported no potential conflict of interest.

Author details

Daniel Nsengumuremyi[1,2,*], Parise Adadi[3], Gavers K. Oppong[4,5], Nadezhda V. Barakova[1] and Elena F. Krivoshapkina[2]

1 Department of Food Biotechnology for Plant Origin Products, ITMO University, St. Petersburg, Russian Federation

2 Institute of Solution Chemistry of Advanced Materials and Technology, ITMO University, St. Petersburg, Russian Federation

3 Department of Food Science, University of Otago, Dunedin, New Zealand

4 Department of Plant Sciences, Rothamsted Research, Hertfordshire, United Kingdom

5 School of Biosciences, University of Nottingham, Loughborough, United Kingdom

*Address all correspondence to: nsedanco@yahoo.fr; nsedanco@gmail.com

References

[1] Narvankar DS, Singh CB, Jayas DS, White NDG. Assessment of soft X-ray imaging for detection of fungal infection in wheat. Biosystems Engineering. 2009;**103**:49-56

[2] Nawrocka A, Cieœla J. Influence of silver nanoparticles on food components in wheat. International Agrophysics. 2013;**27**:49-55

[3] El-Kady IA, Abdel-Hafez SII, El-Maraghy SS. Contribution to the fungal flora of cereal grains in Egypt. Mycopathologia. 1982;**77**:103-109

[4] Kitherian S. Nano and bio-nanoparticles for insect control. Research Journal of Nanoscience and Nanotechnology. 2017;7(1):1-9

[5] Alif AA, Thangapandiyan S. Comparative bioassay of silver nanoparticles and malathion on infestation of red flour beetle, *Tribolium castaneum*. Journal of Basic and Applied Zoology. 2019;**80**:55

[6] Zallen R, Moret M. The optical absorption edge of brookite TiO_2. Solid State Communications. 2006;**137**(3):154-157

[7] Shukla RK, Sharma V, Pandey AK, Singh S, Sultana S, Dhawan A. ROS-mediated genotoxicity induced by titanium dioxide nanoparticles in human epidermal cells. Toxicology In Vitro. 2011;**25**(1):231-241

[8] Morishige T, Yoshioka Y, Tanabe A, Yao X, Tsunoda S-I, Tsutsumi Y, et al. Titanium dioxide induces different levels of IL-1β production dependent on its particle characteristics through caspase-1 activation mediated by reactive oxygen species and cathepsin B. Biochemical and Biophysical Research Communications. 2010;**392**(2):160-165

[9] Macwan DP, Dave PN, Shalini C. A review on nano-TiO_2 sol–gel type syntheses and its applications. Journal of Materials Science. 2011;**46**:3669-3686

[10] Buettner KM, Valentine AM. Bioinorganic chemistry of titanium. Chemical Reviews. 2012;**112**:1863-1881

[11] Brun E, Barreau F, Veronesi G, Fayard B, Sorieul S, Chanéac C, et al. Titanium dioxide nanoparticle impact and translocation through ex vivo, in vivo and in vitro gut epithelia. Particle and Fibre Toxicology. 2014;**11**(1):13

[12] Banerjee K, Thiagarajan P. A review of titanium di oxide nanoparticles—Synthesis, applications and toxicity concerns. Nanoscience Nanotechnology-Asia. 2014;**4**:132-143

[13] Matsunaga T, Tamoda R, Nakajima T, Wake H. Photoelectrochemical sterilization of microbial cells by semiconductor powders. FEMS Microbiology Letters. 1985;**29**:211-214

[14] Rincón A, Pulgarin C. Photocatalytical inactivation of *E. coli*: Effect of (continuous–intermittent) light intensity and of (suspended–fixed) TiO_2 concentration. Applied Catalysis B: Environmental. 2003;**44**(3):263-284

[15] Rincon A. Effect of pH, inorganic ions, organic matter and H_2O_2 on *E. coli* K12 photocatalytic inactivation by TiO_2 implications in solar water disinfection. Applied Catalysis B: Environmental. 2004;**51**(4):283-302

[16] Benabbou A, Derriche Z, Felix C, Lejeune P, Guillard C. Photocatalytic inactivation of *Escherischia coli*: Effect of concentration of TiO_2 and microorganism, nature, and intensity of UV irradiation. Applied Catalysis B: Environmental. 2007;**76**(3-4):257-263

[17] Pigeot-Rémy S, Simonet F, Errazuriz-Cerda E, Lazzaroni J,

Atlan D, Guillard C. Photocatalysis and disinfection of water: Identification of potential bacterial targets. Applied Catalysis B: Environmental. 2011;**104**(3-4):390-398

[18] Pigeot-Rémy S, Simonet F, Atlan D, Lazzaroni J, Guillard C. Bactericidal efficiency and mode of action: A comparative study of photochemistry and photocatalysis. Water Research. 2012;**46**(10):3208-3218

[19] Zimbone M, Buccheri M, Cacciato G, Sanz R, Rappazzo G, Boninelli S, et al. Photocatalytical and antibacterial activity of TiO_2 nanoparticles obtained by laser ablation in water. Applied Catalysis B: Environmental. 2015;**165**:487-494

[20] Xing Y, Li X, Zhang L, Xu Q, Che Z, Li W, et al. Effect of TiO_2 nanoparticles on the antibacterial and physical properties of polyethylene-based film. Progress in Organic Coatings. 2012;**73**(2-3):219-224

[21] Maness PC, Smolinski S, Blake DM, Huang Z, Wolfrum EJ, Jacoby WA. Bactericidal activity of photocatalytic TiO_2 reaction: Toward an understanding of its killing mechanism. Applied and Environmental Microbiology. 1999;**65**:4094-4098

[22] Choi YS, Kim BW. Lactic acid recovery from fermentation broth using one-stage electrodialysis. Journal of Chemical Technology and Biotechnology. 2000;**75**:1145-1150

[23] Wist J, Sanabria J, Dierolf C, Torres W, Pulgarin C. Evaluation of photocatalytic disinfection of crude water for drinking-water production. Journal of Photochemical and Photobiology A: Chemistry. 2002;**147**:241-246

[24] Kim B, Kim D, Cho D, Cho S. Bactericidal effect of TiO_2 photocatalyst on selected food-borne pathogenic bacteria. Chemosphere. 2003;**52**:277-281

[25] Hur JS, Oh SO, Lim KM, Jung JS, Kim JW, Koh YJ. Novel effects of TiO_2 photocatalytic ozonation on control of postharvest fungal spoilage of kiwifruit. Postharvest Biology and Technology. 2005;**35**:109-113

[26] Maneerat C, Hayata Y. Antifungal activity of TiO_2 photocatalysis against *Penicillium expansum* in vitro and in fruit tests. International Journal of Food Microbiology. 2006;**107**:99-103

[27] Kühn KP, Chaberny IF, Massholder K, Stickler M, Benz VW, Sonntag H-G, et al. Disinfection of surfaces by photocatalytic oxidation with titanium dioxide and UVA light. Chemosphere. 2003;**53**(1):71-77

[28] Gupta K, Singh RP, Pandey A, Pandey A. Photocatalytic antibacterial performance of TiO_2 and Ag-doped TiO_2 against *S. aureus*, *P. aeruginosa* and *E. coli*. Beilstein Journal of Nanotechnology. 2013;**4**:345-351

[29] Fekete-Kertész I, Maros G, Gruiz K, Molnár M. The effect of TiO_2 nanoparticles on the aquatic ecosystem: A comparative ecotoxicity study with test organisms of different trophic levels. Periodica Polytechnica, Chemical Engineering. 2016;**60**(4):231-243

[30] Kau JH, Sun DS, Huang HH, Wong MS, Lin HC, Chang HH. Role of visible light-activated photocatalyst on the reduction of anthrax spore-induced mortality in mice. PLoS One. 2009;**4**(1)

[31] Seven O, Dindar B, Aydemir S, Metin D, Ozinel M, Icli S. Solar photocatalytic disinfection of a group of bacteria and fungi aqueous suspensions with TiO_2, ZnO and Sahara Desert dust. Journal of Photochemistry and Photobiology A: Chemistry. 2004;**165**(1-3):103-107

[32] Chen F, Yang X, Wu Q. Antifungal capability of TiO_2 coated film on moist wood. Building and Environment. 2009;**44**(5):1088-1093

[33] Pokhum C, Viboonratanasri D, Chawengkijwanich C. New insight into the disinfection mechanism of *Fusarium monoliforme* and *Aspergillus niger* by TiO_2 photocatalyst under low intensity UVA light. Journal of Photochemistry and Photobiology B: Biology. 2017;**176**:17-24

[34] Wolfrum EJ, Huang J, Blake DM, Maness P-C, Huang Z, Fiest J, et al. Photocatalytic oxidation of bacteria, bacterial and fungal spores, and model biofilm components to carbon dioxide on titanium dioxide-coated surfaces. Environmental Science & Technology. 2002;**36**(15):3412-3419

[35] Erkan A, Bakir U, Karakas G. Photocatalytic microbial inactivation over Pd doped SnO_2 and TiO_2 thin films. Journal of Photochemistry and Photobiology A: Chemistry. 2006;**184**(3):313-321

[36] Anbeaki RAA, Hameed FR, Gassim FAG. Antifungal activity of titanum dioxide photocatalysis against *Fusarium oxysporum* f.sp. lycopersici. Euphrates Journal of Agricultural Science. 2009;**1**:14-26

[37] Lonnen J, Kilvington S, Kehoe SC, Touati FA, McGuigan KG. Solar and photocatalytic disinfection of protozoan, fungal and bacterial microbes in drinking water. Water Research. 2005;**39**(5):877-883. DOI: 10.1016/j.watres.2004.11.023.

[38] Yao Y, Ohko Y, Sekiguchi Y, Fujishima A, Kubota Y. Self-sterilization using silicone catheters coated with Ag and TiO_2 nanocomposite thin film. Journal of Biomedical Materials Research. Part B: Applied Biomaterials. 2008;**85B**(2):453-460. DOI: 10.1002/jbm.b.30965

[39] Veselá M, Veselý M, Chomoucká J, Lipenská M. Photocatalytic disinfection of water using Ag/TiO_2. Chemicke Listy. 2008;**102**:507-508

[40] Sawada D, Ohmasa M, Fukuda M, Masuno K, Koide H, Tsunoda S, et al. Disinfection of some pathogens of mushroom cultivation by photocatalytic treatment. Mycoscience. 2005;**46**(1):54-60. DOI: 10.1007/s10267-004-0211-y

[41] Giannantonio DJ, Kurth JC, Kurtis KE, Sobecky PA. Effects of concrete properties and nutrients on fungal colonization and fouling. International Biodeterioration & Biodegradation. 2009;**63**(3):252-259. DOI: 10.1016/j.ibiod.2008.10.002

[42] Lu JW, Li FB, Guo T, Lin LW, Hou MF, Liu TX. TiO_2 photocatalytic antifungal technique for crops diseases control. Journal of Environmental Sciences. 2006;**18**:397-401

[43] Koide S, Nonami T. Disinfecting efficacy of a plastic container covered with photocatalyst for postharvest. Food Control. 2007;**18**(1):1-4. DOI: 10.1016/j.foodcont.2005.08.001

[44] Sichel C, de Cara M, Tello J, Blanco J, Fernández-Ibáñez P. Solar photocatalytic disinfection of agricultural pathogenic fungi: *Fusarium* species. Applied Catalysis B: Environmental. 2007;**74**(1-2):152-160. DOI: 10.1016/j.apcatb.2007.02.005.

[45] Sichel C, Tello J, de Cara M, Fernández-Ibáñez P. Effect of UV solar intensity and dose on the photocatalytic disinfection of bacteria and fungi. Catalysis Today. 2007;**129**(1-2):152-160. DOI: 10.1016/j.cattod.2007.06.061

[46] Lin CY, Li CS. Effectiveness of titanium dioxide photocatalyst filters for controlling bioaerosols. Aerosol Science and Technology. 2003;**37**(2):162-170. DOI: 10.1080/02786820300951

[47] Lin CY, Li CS. Inactivation of microorganisms on the photocatalytic surfaces in air. Aerosol Science and Technology. 2003;**37**(12):939-946. DOI: 10.1080/02786820300900

[48] Thabet S, Simonet F, Lemaire M, Guillard C, Cotton P. Impact of photocatalysis on fungal cells: Depiction of cellular and molecular effects on *Saccharomyces cerevisiae*. Applied and Environmental Microbiology. 2014;**80**(24):7527-7535

[49] Fujishima A, Rao TN, Tryk DA. Titanium dioxide photocatalysis. Journal of Photochemistry and Photobiology C. 2000;**1**:1-21

[50] Rajeshwar K, Osugi M, Chanmanee W, Chenthamarakshan C, Zanoni M, Kajitvichyanukul P, et al. Heterogeneous photocatalytic treatment of organic dyes in air and aqueous media. Journal of Photochemistry and Photobiology C: Photochemistry Reviews. 2008;**9**(4):171-192

[51] Rehman S, Ullah R, Butt A, Gohar N. Strategies of making TiO_2 and ZnO visible light active. Journal of Hazardous Materials. 2009;**170**(2-3):560-569

[52] Saravanan R, Gracia F, Stephen A. Basic principles, mechanism, and challenges of photocatalysis. In: Khan MM, Pradhan D, Sohn Y, editors. Nanocomposites for Visible Light-induced Photocatalysis. Springer Series on Polymer and Composite Materials. Switzerland: Springer International Publishing; 2017. pp. 19-40

[53] Khan MM, Adil SF, Al-Mayouf A. Metal oxides as photocatalysts. Journal of Saudi Chemical Society. 2015;**19**(5):462-464

[54] Konstantinou IK, Albanis TA. TiO_2-assisted photocatalytic degradation of azo dyes in aqueous solution: kinetic and mechanistic investigations. Applied Catalysis B: Environmental. 2004;**49**(1):1-14

[55] Nakata K, Fujishima A. TiO_2 photocatalysis: Design and applications. Journal of Photochemistry and Photobiology C: Photochemistry Reviews. 2012;**13**(3):169-189

[56] Gogniat G, Dukan S. TiO_2 photocatalysis causes DNA damage via Fenton reaction-generated hydroxyl radicals during the recovery period. Applied and Environmental Microbiology. 2007;**73**(23):7740-7743

[57] Riegel G, Bolton JR. Photocatalytic efficiency variability in TiO_2 particles. The Journal of Physical Chemistry. 1995;**99**(12):4215-4224

[58] Chen B, Le W, Wang Y, Li Z, Wang D, Lin L, et al. Targeting negative surface charges of cancer cells by multifunctional nanoprobes. Theranostics. 2016;**6**(11):1887-1898. DOI: 10.7150/thno.16358

[59] Bishop JR, Schuksz M, Esko JD. Heparan sulphate proteoglycans fine-tune mammalian physiology. Nature. 2007;**446**(7139):1030-1037

[60] Blake DM, Maness PC, Huang Z, Wolfrum EJ, Huang J, Jacoby WA. Application of the photocatalytic chemistry of titanium dioxide to disinfection and the killing of cancer cells. Separation and Purification Methods. 1999;**28**(1):1-50

[61] Youngman RJ. Oxygen activation: is the hydroxyl radical always biologically relevant? Trends in Biochemical Sciences. 1984;**9**(6):280-283

[62] Draper HH, Hadley M. Malondialdehyde determination as index of lipid peroxidation. Methods in Enzymology. 1990;**186**:421-431

[63] Thabet S, Weiss-Gayet M, Dappozze F, Cotton P, Guillard C.

Photocatalysis on yeast cells: Toward targets and mechanisms. Applied Catalysis B: Environmental. 2013;**140-141**:169-178

[64] Schobert M, Gorisch H. Cytochrome C550 is an essential component of the quinoprotein ethanol oxidation system in *Pseudomonas aeruginosa*: Cloning and sequencing of the genes encoding cytochrome C550 and an adjacent acetaldehyde dehydrogenase. Microbiology. 1999;**145**(2):471-481

[65] Kubacka A, Diez MS, Rojo D, Bargiela R, Ciordia S, Zapico I, et al. Understanding the antimicrobial mechanism of TiO_2-based nanocomposite films in a pathogenic bacterium. Scientific Reports. 2014;**4**(1):4134

[66] Dunford R, Salinaro A, Cai L, Serpone N, Horikoshi S, Hidaka H, et al. Chemical oxidation and DNA damage catalysed by inorganic sunscreen ingredients. FEBS Letters. 1997;**418**(1-2):87-90

[67] Kikuchi Y, Sunada K, Iyoda T, Hashimoto K, Fujishima A. Photocatalytic bactericidal effect of TiO_2 thin films: Dynamic view of the active oxygen species responsible for the effect. Journal of Photochemistry and Photobiology A: Chemistry. 1997;**106**(1-3):51-56

[68] Hidaka H, Horikoshi S, Serpone N, Knowland J. In vitro photochemical damage to DNA, RNA and their bases by an inorganic sunscreen agent on exposure to UVA and UVB radiation. Journal of Photochemistry and Photobiology A: Chemistry. 1997;**111**(1-3):205-213

[69] Cezara B, Alexandra C, Janeta T, Svetlana L, Steliana C, Elena D, et al. Effect of nano-oxides TiO_2 and Fe_3O_4 on lipase biosynthesis by *Aspergillus niger* CNMN-FD-01 micromycete. Journal of the Academy of Sciences of Moldova. Life Sciences. 2017;**2**:125-130

[70] Gomes SIL, Roca CP, Kammer FVD, Scott-Fordsmand JJ, Amorim MJB. Mechanisms of (photo) toxicity of TiO_2 nanomaterials (NM103, NM104, NM105): Using high-throughput gene expression in *Enchytraeus crypticus*. Nanoscale. 2018;**10**(46):21960-21970

[71] Dagata A, Fasulo S, Dallas LJ, Fisher AS, Maisano M, Readman JW, et al. Enhanced toxicity of 'bulk titanium dioxide compared to 'fresh and 'aged nano-TiO_2 in marine mussels (*Mytilus galloprovincialis*). Nanotoxicology. 2013;**8**(5):549-558

[72] Blanco-Galvez J, Fernández-Ibáñez P, Malato-Rodríguez S. Solar photocatalytic detoxification and disinfection of water: Recent overview. Journal of Solar Energy Engineering. 2007;**129**(1). DOI: 10.1115/1.2390948

[73] Elieh-Ali-Komi D, Hamblin MR. Chitin and chitosan: Production and application of versatile biomedical nanomaterials. International Journal of Advanced Research. 2016;**4**:411-427

[74] Sudarshan NR, Hoover DG, Knorr D. Antibacterial action of chitosan. Food Biotechnology. 1992;**6**:257-272

[75] Madhumathi K, Kumar PTS, Abhilash S, Sreeja V, Tamura H, Manzoor K, et al. Development of novel chitin/nanosilver composite scaffolds for wound dressing applications. Journal of Materials Science: Materials in Medicine. 2009;**21**(2):807-813

[76] Quan H, Zhu F, Han X, Xu Z, Zhao Y, Miao Z. Mechanism of anti-angiogenic activities of chitooligosaccharides may be through inhibiting heparanase activity. Medical Hypotheses. 2009;**73**(2):205-206

[77] Pangestuti R, Kim S-K. Neuroprotective properties of chitosan and its derivatives. Marine Drugs. 2010;**8**(7):2117-2128

[78] Li Q, Dunn EJ, Grandmaison EW, Goosen MFA. Applications and properties of chitosan. Journal of Bioactive and Compatible Polymers. 1992;7:370-397

[79] Fernandes JC, Tavaria FK, Soares JC, Ramos ÓS, Monteiro MJ, Pintado ME, et al. Antimicrobial effects of chitosans and chitooligosaccharides, upon *Staphylococcus aureus* and *Escherichia coli*, in food model systems. Food Microbiology. 2008;**25**(7):922-928

[80] Wang Y, Zhou P, Yu J, Pan X, Wang P, Lan W, et al. Antimicrobial effect of chitooligosaccharides produced by chitosanase from *Pseudomonas* CUY8. Asia Pacific Journal of Clinical Nutrition. 2007;**16**:174-177

[81] Khanafari A, Marandi R, Sanatei S. Recovery of chitin and chitosan from shrimp waste by chemical and microbial methods. Iranian Journal of Environmental Health Science and Engineering. 2008;**5**:19-24

[82] Li XF, Feng XQ, Yang S. A mechanism of antibacterial activity of chitosan against Gram-negative bacteria. Chinese Journal of Polymer Science. 2010;**31**:148-153

[83] Li XF, Feng XQ, Yang S, Fu GQ, Wang TP, Su ZX. Chitosan kills *Escherichia coli* through damage to be of cell membrane mechanism. Carbohydrate Polymers. 2010b;**79**(3):493-499

[84] Limam Z, Selmi S, Sadok S, El-Abed A. Extraction and characterization of chitin and chitosan from crustacean by-products: Biological and physicochemical properties. African Journal of Biotechnology. 2011;**10**:640-647

[85] Benhabiles M, Salah R, Lounici H, Drouiche N, Goosen M, Mameri N. Antibacterial activity of chitin, chitosan and its oligomers prepared from shrimp shell waste. Food Hydrocolloids. 2012;**29**(1):48-56

[86] Yang E-J, Kim J-G, Kim J-Y, Kim S, Lee N, Hyun C-G. Anti-inflammatory effect of chitosan oligosaccharides in RAW 2647 cells. Open Life Sciences. 2010;**5**(1)

[87] Yu T, Yu C, Chen F, Sheng K, Zhou T, Zunun M, et al. Integrated control of blue mold in pear fruit by combined application of chitosan, a biocontrol yeast and calcium chloride. Postharvest Biology and Technology. 2012;**69**:49-53

[88] Rahman M, Mahmud T, Kadir J, Rahman RA, Begum M. Enhancing the efficacy of *Burkholderia cepacia* B23 with calcium chloride and chitosan to control anthracnose of papaya during storage. The Plant Pathology Journal. 2009;**25**(4):361-368

[89] El-Ghaouth A, Smilanick JL, Wilson CL. Enhancement of the performance of *Candida saitoana* by the addition of glycolchitosan for the control of postharvest decay of apple and citrus fruit. Postharvest Biology and Technology. 2000;**19**(1):103-110

[90] El-Ghaouth A, Smilanick JL, Brown GE, Ippolito A, Wisniewski M, Wilson CL. Application of *Candida saitoana* and glycolchitosan for the control of postharvest diseases of apple and citrus fruit under semi-commercial conditions. Plant Disease. 2000;**84**(3):243-248

[91] Capdeville GD, Wilson CL, Beer SV, Aist JR. Alternative disease control agents induce resistance to blue mold in harvested 'red delicious' apple fruit. Phytopathology. 2002;**92**(8):900-908

[92] Kaur P, Thakur R, Barnela M, Chopra M, Manuja A, Chaudhury A. Synthesis, characterization and in vitro evaluation of cytotoxicity and antimicrobial activity of chitosan-metal

nanocomposites. Journal of Chemical Technology & Biotechnology. 2014;**90**(5):867-873

[93] Hashim AF, Alghuthaymi MA, Vasil'kov AY, Abd-Elsalam KA. Polymer inorganic nanocomposites: A sustainable antimicrobial agents. In: Prasad R, editor. Advances and Applications through Fungal Nanobiotechnology. Switzerland: Springer International Publishing; 2016. pp. 265-290

[94] Brunel F, Gueddari NEE, Moerschbacher BM. Complexation of copper(II) with chitosan nanogels: Toward control of microbial growth. Carbohydrate Polymers. 2013;**92**(2):1348-1356

[95] Vokhidova NR, Sattarov ME, Kareva ND, Rashidova SS. Fungicide features of the nanosystems of silkworm (*Bombyx mori*) chitosan with copper ions. Microbiology. 2014;**83**(6):751-753

[96] Sagoo S. The antimicrobial action of chitosan [PhD thesis]. Newington: South Bank University; 2003

[97] Sagoo S, Board R, Roller S. Chitosan potentiates the antimicrobial action of sodium benzoate on spoilage yeasts. Letters in Applied Microbiology. 2002;**34**(3):168-172

[98] Pospieszny H, Chirkov S, Atabekov J. Induction of antiviral resistance in plants by chitosan. Plant Science. 1991;**79**(1):63-68

[99] Kulikov SN, Chirkov SN, Il'Ina AV, Lopatin SA, Varlamov VP. Effect of the molecular weight of chitosan on its antiviral activity in plants. Applied Biochemistry and Microbiology. 2006;**42**(2):200-203

[100] Yang F-L, Li X-G, Zhu F, Lei C-L. Structural characterization of nanoparticles loaded with garlic essential oil and their insecticidal activity against *Tribolium castaneum* (Herbst) (Coleoptera: Tenebrionidae). Journal of Agricultural and Food Chemistry. 2009;**57**(21):10156-10162

[101] Mgbeahuruike E, Fyhrquist P, Vuorela H, Julkunen-Tiitto R, Holm Y. Alkaloid-rich crude extracts, fractions and piperamide alkaloids of piper guineense possess promising antibacterial effects. Antibiotics. 2018;7(4):98

[102] Nychas GJE, Skandamis PN. Antimicrobials from herbs and spices. In: Roller S, editor. Natural Antimicrobials for the Minimal Processing of Foods. Cambridge: Woodhead Publishing Limited; 2000. pp. 176-192

[103] Ippolito A, Nigro F. Natural antimicrobials in postharvest storage of fresh fruits and vegetables. In: Roller S, editor. Natural Antimicrobials for the Minimal Processing of Foods. Cambridge, UK: Woodhead Publishing Limited; 2000. pp. 201-224

[104] Fang SW, Li CF, Shih DYC. Antifungal activity of chitosan and its preservative effect on low-sugar candied kumquat. Journal of Food Protection. 1994;**57**(2):136-140. DOI: 10.4315/0362-028x-57.2.136

[105] Roller S, Covill N. The antifungal properties of chitosan in laboratory media and apple juice. International Journal of Food Microbiology. 1999;**47**(1-2):67-77

[106] Badawy MEI, Ahmed M, Rabea EI. Bactericidal and fungicidal activities of different molecular weight chitosan samples. Pest Control. 2006;**14**:19-34

[107] Martinez LR, Mihu MR, Han G, Frases S, Cordero RJ, Casadevall A, et al. The use of chitosan to damage *Cryptococcus neoformans* biofilms. Biomaterials. 2010;**31**(4):669-679. DOI: 10.1016/j.biomaterials.2009.09.087

[108] Jiang T, Feng L, Li J. Changes in microbial and postharvest quality of shiitake mushroom (*Lentinus edodes*) treated with chitosan-glucose complex coating under cold storage. Food Chemistry. 2012;**131**(3):780-786. DOI: 10.1016/j.foodchem.2011.08.087

[109] Tikhonov VE, Stepnova EA, Babak VG, Yamskov IA, Palma-Guerrero J, Jansson HB, et al. Bactericidal and antifungal activities of a low molecular weight chitosan and its N-/2(3)-(dodec-2-enyl)succinoyl/-derivatives. Carbohydrate Polymers. 2006;**64**(1):66-72

[110] Ramos-García M, Bosquez-Molina E, Hernández-Romano J, Zavala-Padilla G, Terrés-Rojas E, Alia-Tejacal I, et al. Use of chitosan-based edible coatings in combination with other natural compounds, to control *Rhizopus stolonifer* and *Escherichia coli* DH5α in fresh tomatoes. Crop Protection. 2012;**38**:1-6. DOI: 10.1016/j.cropro.2012.02.016

[111] Ziani K, Fernández-Pan I, Royo M, Maté JI. Antifungal activity of films and solutions based on chitosan against typical seed fungi. Food Hydrocolloids. 2009;**23**(8):2309-2314. DOI: 10.1016/j.foodhyd.2009.06.005

[112] Vu KD, Hollingsworth RG, Leroux E, Salmieri S, Lacroix M. Development of edible bioactive coating based on modified chitosan for increasing the shelf life of strawberries. Food Research International. 2011;**44**(1):198-203. DOI: 10.1016/j.foodres.2010.10.037

[113] Zhang M, Tan T, Yuan H, Rui C. Insecticidal and fungicidal activities of chitosan and oligo-chitosan. Journal of Bioactive and Compatible Polymers. 2003;**18**(5):391-400. DOI: 10.1177/0883911503039019

[114] Sebti I, Martial-Gros A, Carnet-Pantiez A, Grelier S, Coma V. Chitosan polymer as bioactive coating and film against *Aspergillus niger* contamination. Journal of Food Science. 2005;**70**(2):M100–M104. DOI: 10.1111/j.1365-2621.2005.tb07098.x

[115] Bhaskara Reddy MV, Arul J, Ait-Barka E, Angers P, Richard C, Castaigne F. Effect of chitosan on growth and toxin production by *Alternaria alternata* f. sp. lycopersici. Biocontrol Science and Technology. 1998;**8**(1):33-43. DOI: 10.21273/hortsci.32.3.467f

[116] Dash M, Chiellini F, Ottenbrite R, Chiellini E. Chitosan—A versatile semi-synthetic polymer in biomedical applications. Progress in Polymer Science. 2011;**36**(8):981-1014

[117] Ifuku S. Chitin and chitosan nanofibers: Preparation and chemical modifications. Molecules. 2014;**19**(11):18367-18380

[118] Xing K, Chen XG, Kong M, Liu CS, Cha DS, Park HJ. Effect of oleoyl-chitosan nanoparticles as a novel antibacterial dispersion system on viability, membrane permeability and cell morphology of *Escherichia coli* and *Staphylococcus aureus*. Carbohydrate Polymers. 2009;**76**(1):17-22

[119] Xing K, Chen XG, Liu CS, Cha DS, Park HJ. Oleoyl-chitosan nanoparticles inhibits *Escherichia coli* and *Staphylococcus aureus* by damaging the cell membrane and putative binding to extracellular or intracellular targets. International Journal of Food Microbiology. 2009;**132**(2-3):127-133

[120] Kong M, Chen XG, Xing K, Park HJ. Antimicrobial properties of chitosan and mode of action: A state of the art review. International Journal of Food Microbiology. 2010;**144**(1):51-63

[121] Lin G, Zhang Q, Lin X, Zhao D, Jia R, Gao N, et al. Enhanced photoluminescence of gallium phosphide by surface plasmon resonances of metallic nanoparticles. RSC Advances. 2015;**5**(60):48275-48280

[122] Chien RC, Yen MT, Mau JL. Antimicrobial and antitumor activities of chitosan from shiitake stipes, compared to commercial chitosan from crab shells. Carbohydrate Polymers. 2016;**138**:259-264

[123] Severino R, Ferrari G, Vu KD, Donsì F, Salmieri S, Lacroix M. Antimicrobial effects of modified chitosan based coating containing nanoemulsion of essential oils, modified atmosphere packaging and gamma irradiation against *Escherichia coli* O157:H7 and *Salmonella Typhimurium* on green beans. Food Control. 2015;**50**:215-222

[124] Yang TC, Chou CC, Li CF. Antibacterial activity of N-alkylated disaccharide chitosan derivatives. International Journal of Food Microbiology. 2005;**97**(3):237-245

[125] Ignatova M, Starbova K, Markova N, Manolova N, Rashkov I. Electrospun nano-fibre mats with antibacterial properties from quaternised chitosan and polyvinyl alcohol. Carbohydrate Research. 2006;**341**(12):2098-2107

[126] Jia Z, Shen D, Xu W. Synthesis and antibacterial activities of quaternary ammonium salt of chitosan. Carbohydrate Research. 2001;**333**(1):1-6

[127] Xie Y, Liu X, Chen Q. Synthesis and characterization of water-soluble chitosan derivate and its antibacterial activity. Carbohydrate Polymers. 2007;**69**(1):142-147

[128] Qin C, Li H, Xiao Q, Liu Y, Zhu J, Du Y. Water-solubility of chitosan and its antimicrobial activity. Carbohydrate Polymers. 2006;**63**(3):367-374

[129] Li Z, Yang F, Yang R. Synthesis and characterization of chitosan derivatives with dual-antibacterial functional groups. International Journal of Biological Macromolecules. 2015;**75**:378-387

[130] Márquez IG, Akuaku J, Cruz I, Cheetham J, Golshani A, Smith ML. Disruption of protein synthesis as antifungal mode of action by chitosan. International Journal of Food Microbiology. 2013;**164**(1):108-112

[131] Ing LY, Zin NM, Sarwar A, Katas H. Antifungal activity of chitosan nanoparticles and correlation with their physical properties. International Journal of Biomaterials. 2012:1-9

[132] Raafat D, Sahl H-G. Chitosan and its antimicrobial potential—A critical literature survey. Microbial Biotechnology. 2009;**2**(2):186-201

[133] Helander IM, Nurmiaho-Lassila EL, Ahvenainen R, Rhoades J, Roller S. Chitosan disrupts the barrier properties of the outer membrane of Gram-negative bacteria. International Journal of Food Microbiology. 2001;**71**(2-3):235-244

[134] Zakrzewska A, Boorsma A, Brul S, Hellingwerf KJ, Klis FM. Transcriptional response of *Saccharomyces cerevisiae* to the plasma membrane-perturbing compound chitosan. Eukaryotic Cell. 2005;**4**(4):703-715

[135] Je JY, Kim SK. Chitosan derivatives killed bacteria by disrupting the outer and inner membrane. Journal of Agricultural and Food Chemistry. 2006;**54**(18):6629-6633

[136] Torr KM, Chittenden C, Franich RA, Kreber B. Advances in understanding bioactivity of chitosan and chitosan oligomers against selected wood-inhabiting fungi. Holzforschung. 2005;**59**(5):559-567

[137] Helander IM, Wright AV, Mattila-Sandholm TM. Potential of lactic acid bacteria and novel antimicrobials against Gram-negative bacteria. Trends in Food Science & Technology. 1997;**8**(5):146-150

[138] Kong M, Chen XG, Liu CS, Liu CG, Meng XH, Yu LJ. Antibacterial mechanism of chitosan microspheres in a solid dispersing system against *E. coli*. Colloids and Surfaces B: Biointerfaces. 2008;**65**(2):197-202

[139] Raafat D, Bargen KV, Haas A, Sahl H-G. Insights into the mode of action of chitosan as an antibacterial compound. Applied and Environmental Microbiology. 2008;**74**(23):3764-3773

[140] Rabea EI, Badawy ME-T, Stevens CV, Smagghe G, Steurbaut W. Chitosan as antimicrobial agent: Applications and mode of action. Biomacromolecules. 2003;**4**(6): 1457-1465

[141] Goy RC, Britto D, Assis OBG. A review of the antimicrobial activity of chitosan. Polímeros. 2009;**19**:241-247

[142] Sahariah P, Másson M. Antimicrobial chitosan and chitosan derivatives: A review of the structure–activity relationship. Biomacromolecules. 2017;**18**(11): 3846-3868

[143] Palma-Guerrero J, Lopez-Jimenez J, Pérez-Berná A, Huang I-C, Jansson HB, Salinas J, et al. Membrane fluidity determines sensitivity of filamentous fungi to chitosan. Molecular Microbiology. 2010;**75**:1021-1032

[144] Hadwiger LA, Beckman JM, Adams MJ. Localization of fungal components in the pea-*Fusarium* interaction detected immunochemically with anti-chitosan and anti-fungal cell wall antisera. Plant Physiology. 1981;**67**(1):170-175

[145] Tarsi R, Muzzarelli R, Guzmàn C, Pruzzo C. Inhibition of *Streptococcus mutans* adsorption to hydroxyapatite by low-molecular-weight chitosans. Journal of Dental Research. 1997;**76**(2):665-672

[146] Liu XF, Guan YL, Yang DZ, Li Z, Yao KD. Antibacterial action of chitosan and carboxymethylated chitosan. Journal of Applied Polymer Science. 2000;**79**(7):1324-1335

[147] Hosseinnejad M, Jafari SM. Evaluation of different factors affecting antimicrobial properties of chitosan. International Journal of Biological Macromolecules. 2016;**85**:467-475

[148] Barakova NV, Sharova NY, Juskauskajte AR, Mityukov AS, Romanov VA, Nsengumuremyi D. Fungicidal activity of ultradisperse humic sapropel suspensions. Agronomy Research. 2017;**15**(3):639-648

[149] Avdeyeva LN, Kovalyenko TA, Krivonos OI, Plaksin GV, Strunina NN. Determination of sapropel chemical composition. Chemistry and Chemical Engineering. 2009;**52**:121-123 (in Russian)

[150] Nsengumuremyi D, Barakova NV, Romanov VA, Guzeva AV. The effect of sapropel extracts on microflora and physicochemical parameters of dried distillers' grain. Agronomy Research. 2018;**16**(S2):1457-1465

[151] Nsengumuremyi D, Adadi P, Ukolova MV, Barakova NV. Effects of ultradisperse humic sapropel suspension on microbial growth and fermentation parameters of barley distillate. Fermentation. 2019;**5**:24

[152] Kireycheva LV, Khokhlova OB. Elemental composition of humic substances insapropel deposits. RAAS Newsletter. 2000;**4**:59-62 (in Russian)

[153] Gorbunovskaya OM, Kurzo VB. New methods of sapropel composition analysis. Chemistry and Chemical Engineering. 2001;**2**:73-81 (in Russian)

[154] Dolgopolov VN. Trials of Gumival™ yield better bovine, swine and fowl stock productivity. In: Proceedings of Conference on Summaries and Perspectives of Humin

Preparations in Livestock Industry; Moscow. 2006. pp. 40-43 (in Russian)

[155] Kulikova NA, Filippova OI, Abros'kin DP, Klyain OI. Modern approach to humic substance biological activity assessment. In: Proceedings of International Conference on Biodiagnostics in Soil Ecology Analysis. Moscow: BINOM; 2013. pp. 116-111 (in Russian)

[156] Platonov VV, Khadartsev AA, Fridzon KY, Chunosov SN. Chemical composition and biological activity of sapropel from the Lake Glubokoe (Tatarstan). Journal of New Medical Technologies. 2014;**21**(3):199-204 (in Russian)

[157] Rumyantsev VA, Mityukov AS, Kryukov LN, Yaroshevich GS. Unique properties of humic substances from sapropel. Doklady Earth Sciences. 2017;**473**(2):482-484

[158] Shtin SM. Lake Sapropels Complex Utilization. Moscow: MSU; 2005 (in Russian)

[159] Kosov VI. Sapropel Resources, Engineering, Geoecology. Saint Petersburg: Nauka; 2007. p. 244 (in Russian)

[160] Schepetkin I, Khlebnikov A, Kwon BS. Medical drugs from humus matter: Focus on mumie. Drug Development Research. 2002;**57**(3):140-159

[161] Buzlama AV, Chernov YN. Pharmacological properties, action pathways and prospectives of humic substances in medicine: An analysis. Russian Journal of Experimental and Clinical Pharmacology. 2010;**73**:43-48 (in Russian)

[162] Ouyang K, Walker SL, Yu XY, Gao CH, Huanga Q, Cai P. Metabolism, survival, and gene expression of *Pseudomonas putida* to hematite nanoparticles mediated by surface-bound humic acid. Environmental Science: Nano. 2018;**5**:682-695

[163] Wu M, Song M, Liu M, Jiang C, Li Z. Fungicidal activities of soil humic/fulvic acids as related to their chemical structures in greenhouse vegetable fields with cultivation chronosequence. Scientific Reports. 2016;**6**(1):32858

[164] Ong KJ, Felix LC, Boyle D, Ede JD, Ma G, Veinot JG, et al. Humic acid ameliorates nanoparticle-induced developmental toxicity in zebrafish. Environmental Science: Nano. 2017;**4**(1):127-137

[165] Ezhkov VO, Yapparov AK, Ezhkova AM, Yapparov IA, Ezhkova GO, Faizrakhmanov RN, et al. Study of the action of nanostructured sapropel at different doses on the morphological and functional condition of the gastrointestinal tract in albino mice. Nanotechnologies in Russia. 2016;**11**(7-8):497-505

[166] Perdue EM. Chemical composition, structure, and metal binding properties. In: Hessen DO, Tranvik L, editors. Aquatic Humic Substances: Ecology and Biogeochemistry. Berlin: Springer-Verlag; 1998. pp. 41-61

[167] Cosgrove DJ. Creeping, walls, softening fruit, and penetrating pollen tubes: The growing role of expansins. Proceedings of the National Academy of Sciences of the United States of America. 1997;**94**:5504-5505

[168] Lipke PN, Ovalle R. Cell wall architecture in yeast: New structure and new challenges. Journal of Bacteriology. 1998;**15**:3735-3740

[169] Gubbins PO, Anaissie EJ. Antifungal therapy. In: Clinical Mycology. 2nd ed. 2009. pp. 161-195

[170] Bittner M. Direct effects of humic substances on organisms [Master's thesis]. Brno: Masaryk University; 2006

[171] Lofts S, Tipping E, Sanchez A, Dodd B. Modelling the role of humic acid in radiocaesium distribution in a British upland peat soil. Journal of Environmental Radioactivity. 2002;**61**(2):133-147. DOI: 10.1016/S0265-931X(01)00118-7

[172] García-Mina J, Antolín M, Sanchez-Diaz M. Metal-humic complexes and plant micronutrient uptake: A study based on different plant species cultivated in diverse soil types. Plant and Soil. 2004;**258**(1):57-68

[173] Heil CA. Influence of humic, fulvic and hydrophilic acids on the growth, photosynthesis and respiration of the dinoflagellate *Prorocentrum minimum* (Pavillard) Schiller. Harmful Algae. 2005;**4**(3):603-618

[174] Schulten H-R, Leinweber P. New insights into organic-mineral particles: Composition, properties and models of molecular structure. Biology and Fertility of Soils. 2000;**30**(5-6):399-432

[175] Wang W-H, Ray CM, Jones MN. The fate of ^{14}C-labelled hurnic substances in rice cells in culture. Journal of Plant Physiology. 1999;**154**(2):203-211

[176] Steinberg CEW, Höss S, Brüggemann R. Further evidence that humic substances have the potential to modulate the reproduction of the nematode *Caenorhabditis elegans*. International Review of Hydrobiology. 2002;**87**(1):121

[177] Vigneault B, Percot A, Lafleur M, Campbell PGC. Permeability changes in model and phytoplankton membranes in the presence of aquatic humic substances. Environmental Science & Technology. 2000;**34**(18):3907-3913

[178] Almatov KT, Akhmerov AI. Mumie effect on oxidative phosphorylation and enzymes of mitochondrial respiratory chain of rat liver, small and large intestine. Tashkent State Medical Institute. 1977;**7**:11-14

[179] Ghosal S. Shilajit. Part 15. Shilajit: Its origin and vital significance. In: Mukerjee B, editor. Traditional Medicine. New Delhi: Oxford & IBH; 1993. pp. 308-319

[180] Bhattacharya SK, Sen AP, Ghosal S. Effects of shilajit on biogenic free radicals. Phytotherapy Research. 1995;**9**(1):56-59

[181] Ghosal S, Tripathi VK, Chauhan S. Active constituents of *Emblica officinalis*: Part 1—The chemistry and antioxidative effects of two new hydrolysable tannins, Emblicanin A and B. Indian Journal of Chemistry: Section B Organic and Medicinal Chemistry. 1996;**35**:941-948

[182] Quaggiotti S. Effect of low molecular size humic substances on nitrate uptake and expression of genes involved in nitrate transport in maize (*Zea mays* L.). Journal of Experimental Botany. 2004;**55**(398):803-813

[183] Visser S. Effect of humic substances on mitochondrial respiration and oxidative phosphorylation. Science of The Total Environment. 1987;**62**:347-354

[184] Lu F-J, Tseng S-N, Li M-L, Shih S-R. In vitro anti-influenza virus activity of synthetic humate analogues derived from protocatechuic acid. Archives of Virology. 2002;**147**(2):273-284

[185] Bittner M, Janošek J, Hilscherová K, Giesy J, Holoubek I, Bláha L. Activation of Ah receptor by pure humic acids. Environmental Toxicology. 2006;**21**(4):338-342

[186] Vallini G, Pera A, Avio L, Valdrighi M, Giovannetti M. Influence of humic acids on laurel growth, associated rhizospheric microorganisms, and mycorrhizal

fungi. Biology and Fertility of Soils. 1993;**16**(1):1-4

[187] Höss S, Bergtold M, Haitzer M, Traunspurger W, Steinberg CE. Refractory dissolved organic matter can influence the reproduction of *Caenorhabditis elegans* (Nematoda). Freshwater Biology. 2001;**46**(1):1-10

[188] Meinelt T, Schreckenbach K, Knopf K, Wienke A, StuBer A, Steinberg CEW. Humic substances affect physiological condition and sex ratio of swordtail (*Xiphophorus helleri* Heckel). Aquatic Sciences. 2004;**66**(2):239-245. DOI: 10.1007/s00027-004-0706-9

[189] Timofeyev MA, Wiegand C, Burnison BK, Shatilina ZM, Pflugmacher S, Steinberg CE. Impact of natural organic matter (NOM) on freshwater amphipods. Science of The Total Environment. 2004;**319**(1-3):115-121

[190] Kang S, Herzberg M, Rodrigues DF, Elimelech M. Antibacterial effects of carbon nanotubes: Size does matter! Langmuir. 2008;**24**(13):6409-6413

[191] Buzea C, Pacheco II, Robbie K. Nanomaterials and nanoparticles: Sources and toxicity. Biointerphases. 2007;**2**(4):MR17–MR71

[192] Cataldo F, Da Ros T. Medicinal Chemistry and Pharmacological Potential of Fullerenes and Carbon Nanotubes. Trieste: Springer; 2008

[193] Wang JT, Chen C, Wang E, Kawazoe Y. A new carbon allotrope with six-fold helical chains in all-sp2 bonding networks. Scientific Report. 2014;**4**:4339

[194] Sokolov VI, Stankevich IV. The fullerenes-new allotropic forms of carbon: Molecular and electronic structure, and chemical properties. Russian Chemical Reviews. 1993;**62**(5):419-435

[195] Gibson N, Shenderova O, Luo TJM, Moseenkov S, Bondar V, Puzyr A, et al. Colloidal stability of modified nanodiamond particles. Diamond and Related Materials. 2009;**18**:620-626

[196] Maleki Dizaj S, Mennati A, Jafari S, Khezri K, Adibkia K. Antimicrobial activity of carbon-based nanoparticles. Advanced Pharmaceutical Bulletin. 2015;**5**(1):19-23. DOI: 10.5681/apb.2015.003

[197] Kovač T, Borišev I, Crevar B, Kenjeri FC, Kovač M, Strelec I, et al. Fullerol $C_{60}(OH)_{24}$ nanoparticles modulate aflatoxin B1 biosynthesis in *Aspergillus flavus*. Scientific Reports. 2018;**8**:12855. DOI: 10.1038/s41598-018-31305-9

[198] Hao Y, Cao X, Ma C, Zhang Z, Zhao N, Ali A, et al. Potential applications and antifungal activities of engineered nanomaterials against Gray mold disease agent *Botrytis cinerea* on rose petals. Frontiers in Plant Science. 2017;**8**:1332. DOI: 10.3389/fpls.2017.01332

[199] Sawangphruk M, Srimuk P, Chiochan P, Sangsri T, Siwayaprahm P. Synthesis and antifungal activity of reduced graphene oxide nanosheets. Carbon. 2012;**50**(14):5156-5516. DOI: 10.1016/j.carbon.2012.06.056

[200] Hudler GW. Magical Mushrooms Mischievous Molds. Vol. 7. Princeton, NJ: Princeton University; 1998. p. 7

[201] Wang X, Liu X, Chen J, Han H, Yuan Z. Evaluation and mechanism of antifungal effects of carbon nanomaterials in controlling plant fungal pathogen. Carbon. 2014;**68**:798-806. DOI: 10.1016/j.carbon.2013.11.072

[202] Wang X, Zhou Z, Chen F. Surface modification of carbon nanotubes with an enhanced antifungal activity for the control of plant fungal pathogen.

Materials. 2017;**10**:1375. DOI: 10.3390/ma10121375

[203] Canuto de Menezes BR, Rodrigues KF, da Silva Fonseca BC, Ribas RG, do Amaral Montanheiro TL, Gilmar Patrocínio Thim GP. Recent advances in the use of carbon nanotubes as smart biomaterials. Journal of Materials Chemistry B. 2019;**7**:1343-1360. DOI: 10.1039/C8TB02419G

[204] Klumpp C, Kostarelos K, Prato M, Bianco A. Functionalized carbon nanotubes as emerging nanovectors for the delivery of therapeutics. Biochimica et Biophysica Acta (BBA) – Biomembranes. 2006;**1758**(3):404-412. DOI: 10.1016/j.bbamem.2005.10.008

[205] Panchakarla LS, Govindaraj A. Covalent and non-covalent functionalization and solubilization of double-walled carbon nanotubes in nonpolar and aqueous media. Journal of Chemical Sciences. 2008;**120**(6):607-611

[206] Zare-Zardini H, Amiri A, Shanbedi M, Memarpoor-Yazdi M, Asoodeh A. Studying of antifungal activity of functionalized multiwalled carbon nanotubes by microwave-assisted technique. Surface and Interface Analysis. 2013;**45**:751-755. DOI: 10.1002/sia.5152

[207] Katerine I, Romina AA, Jorge ES, Natalia B, José RVB, Carlos RG, et al. Antifungal activity of cotton fabrics finished modified silica-silver carbon-based hybrid nanoparticles. Textile Research Journal. 2019;**89**(5):825-833. DOI: 10.1177/0040517518755792

[208] Yuan X, Zhang X, Sun L, Wei Y, Wei X. Cellular toxicity and immunological effects of carbon-based nanomaterials. Particle and Fibre Toxicology. 2019;**16**:18. DOI: 10.1186/s12989-019-0299-z

[209] Saha D, Heldt CL, Gencoglu MF, Vijayaragavan KS, Chen J, Saksule A. A study on the cytotoxicity of carbon-based materials. Materials Science and Engineering C. 2016;**68**:101-108. DOI: 10.1016/j.msec.2016.05.094

[210] Al-Jumaili A, Alancherry S, Bazaka K, Jacob MV. Review on the antimicrobial properties of carbon nanostructures. Materials. 2017;**10**:1066. DOI: 10.3390/ma10091066

[211] Gurunathan S, Han JW, Dayem AA, Eppakayala V, Kim JH. Oxidative stress-mediated antibacterial activity of graphene oxide and reduced graphene oxide in *Pseudomonas aeruginosa*. International Journal of Nanomedicine. 2012;**7**:5901-5914. DOI: 10.2147/IJN.S37397

[212] Krishnamoorthy K, Veerapandian M, Zhang LH, Yun K, Kim SJ. Antibacterial efficiency of graphene nanosheets against pathogenic bacteria via lipid peroxidation. Journal of Physical Chemistry C. 2012;**116**:17280-17287. DOI: 10.1021/jp3047054

[213] Higashisaka K, Yoshioka Y, Tsutsumi Y. Applications and safety of nanomaterials used in the food industry. Food Safety. 2015;**3**(2):39-47

[214] Zande MVD, Vandebriel RJ, Doren EV, Kramer E, Rivera ZH, Serrano-Rojero CS, et al. Distribution, elimination, and toxicity of silver nanoparticles and silver ions in rats after 28-day oral exposure. ACS Nano. 2012;**6**(8):7427-7442

[215] Guan M, Zhu QL, Liu Y, Bei YY, Gu ZL, Zhang XN, et al. Uptake and transport of a novel anticancer drug-delivery system: Lactosyl-norcantharidin-associated N-trimethyl chitosan nanoparticles across intestinal Caco-2 cell monolayers. International Journal of Nanomedicine. 2012;**7**:1921-1930

[216] Saremi S, Dinarvand R, Kebriaeezadeh A, Ostad SN, Atyabi F. Enhanced oral delivery of docetaxel using thiolated chitosan nanoparticles: preparation, in vitro and in vivo studies. BioMed Research International. 2013:1-8

[217] He B, Lin P, Jia Z, Du W, Qu W, Yuan L, et al. The transport mechanisms of polymer nanoparticles in Caco-2 epithelial cells. Biomaterials. 2013;**34**(25):6082-6098

[218] Geiser M, Rothen-Rutishauser B, Kapp N, Schürch S, Kreyling W, Schulz H, et al. Ultrafine particles cross cellular membranes by nonphagocytic mechanisms in lungs and in cultured cells. Environmental Health Perspectives. 2005;**113**(11):1555-1560

[219] He C, Hu Y, Yin L, Tang C, Yin C. Effects of particle size and surface charge on cellular uptake and biodistribution of polymeric nanoparticles. Biomaterials. 2010;**31**(13):3657-3666

[220] Kim YS, Song MY, Park JD, Song KS, Ryu HR, Chung YH, et al. Subchronic oral toxicity of silver nanoparticles. Particle Fibre Toxicology. 2010;**7**:20

[221] Wang J, Zhou G, Chen C, Yu H, Wang T, Ma Y, et al. Acute toxicity and biodistribution of different sized titanium dioxide particles in mice after oral administration. Toxicology Letters. 2007;**168**(2):176-185

[222] Warheit D, Brown S, Donner E. Acute and subchronic oral toxicity studies in rats with nanoscale and pigment grade titanium dioxide particles. Food and Chemical Toxicology. 2015;**84**:208-224

[223] Sharma V, Singh P, Pandey AK, Dhawan A. Induction of oxidative stress, DNA damage and apoptosis in mouse liver after sub-acute oral exposure to zinc oxide nanoparticles. Mutation Research/Genetic Toxicology and Environmental Mutagenesis. 2012;**745**(1-2):84-91

[224] Cho WS, Kang BC, Lee JK, Jeong J, Che JH, Seok SH. Comparative absorption, distribution, and excretion of titanium dioxide and zinc oxide nanoparticles after repeated oral administration. Particle and Fibre Toxicology. 2013;**10**(1):9